Gilberto Ledesma

Volumetric Chemical Analysis of Natural Waters

Gilberto Ledesma

Volumetric Chemical Analysis of Natural Waters

Surface and Underground

ScienciaScripts

Imprint

Cover image: www.ingimage.com

This book is a translation from the original published under ISBN 978-613-9-43676-7.

Publisher:
Sciencia Scripts
is a trademark of
Dodo Books Indian Ocean Ltd. and OmniScriptum S.R.L publishing group

120 High Road, East Finchley, London, N2 9ED, United Kingdom
Str. Armeneasca 28/1, office 1, Chisinau MD-2012, Republic of Moldova, Europe
Printed at: see last page
ISBN: 978-620-8-19351-5

Contents

With a special dedication to my family for always being here, for being present, for being a team that generates ideas: To my parents Gilberto and Elena, my sister Sandra and my dear niece Karol Milena".

"To my partner Genessis for being part of this little universe we have formed".

"To my friends of exploration, for those kilometres and years of life travelled both above and below the surface ...., always with a great desire to understand this world that surrounds us... "

PROLOGUE

Nowadays, the issue of water is something that should be priceless. It is a natural resource that little by little is generating various problems because every day it is becoming more complicated to have access to it. The water problem is real and it is there, every day the levels of lakes, streams and springs are diminishing due to thousands of factors. The issue of water should always be a priority, we abuse it without realising it and the time may come when this vital resource for human development will be increasingly difficult to reach our homes.

The knowledge and distribution of water is fundamental, if we do not know its principles and laws, we can hardly understand it, let alone try to take care of it. Today we need more and more people to take an interest in this increasingly scarce resource. We need a policy that teaches us that the duality of water-human-environment is as vital as water itself.

I present this basic manual of natural water analysis as a contribution to an uncertain future, to an unclear future where understanding what happens to water will be a strategic factor for populations. It is a manual that allows us to approach the understanding and properties of natural water, as we find it in our environment . It is not a manual for water and treated waste . It is a manual that compiles the

The technical techniques and the basic principles of analysis of the

traditional Qrnmica documents and writings

The volumetry in its maximum expression and from a micro-scale point of view, as a simple, cheap and precise anaHtic resource. Volumetry in its maximum expression and from a micro-scale point of view, as a simple, cheap and precise anaHtic resource. A manual to reach to the most remote and unexplored places to take a water sample and, if possible, to carry out a series of tests "in situ" that will allow us to know the most important properties of the water and to plan and analyse actions for the improvement, care, protection and possibly distribution of the vital liquid.

It is the central idea of this small manual; to bring this methodology of anaHthematic qrnmics to people interested in the subject of water, from qrnmics, hydrologists, environmental engineers, explorers and all those who need water data, for its classification and that through the new existing hydroqrnmics programs, we have at hand the context of that natural water, surface and groundwater, which may be in the immediate future, fundamental for simply... ...our existence.

Gilberto Ledesma Ledesma

CHAPTER 1

Water and its hydrological cycle.

Water resources of the earth.

The Earth is the only planet in the solar system with almost 71% of its surface covered with water. The volumes of water on Earth are large. The oceans and seas alone are estimated to contain 1360 x 10^6 km^3 and an estimated 29 x 10^6 km^3 are confined in the polar ice caps. Together these water masses account for approximately 97% of the world's water resources. Unfortunately, none of these volumes are immediately available to supply water, the seawater because it has almost 35 grams of salt per litre and the polar ice because of its remoteness from the habitable regions of the planet.

This leaves as the useful waters of the earth the " 7 x 10^5 km^3 of fresh water in lakes and streams, in permeable soils and in the atmosphere; just under 3% is in the atmosphere, derived from the evaporation of fresh and salt water, the remainder being divided between the surface and subsoil. However, these portions of the hydrosphere are not static, the water circulates: 50 x 10^4 km^3 precipitates annually from the sky in the form of rain or snow on the continents and islands about a quarter; the remainder falls into the oceans; 375 x 10^4 km^3 drains from the earth's surface: and the subsoil yields a more or less safe 1.67 x 10^4 km^3 of water. Table 1.1 shows the uneven distribution of water on Earth.

Place	%
Oceans Polar ice caps, icebergs, glaciers Soil moisture Lakes and rivers Atmosphere	97.39 2.01 0.58 0.02 0.001

Table 1.1 Percentage distribution of land water.

Inland waters.

Table 1.1 shows that approximately 2.6% of the world's water is non-saline or "fresh" water and of this only 10% is found in lakes, rivers and subsoil; this value indicates the water available on the continents. Fresh water is therefore an extremely scarce resource. Freshwater comes from two sources: surface water originating from precipitation that does not infiltrate into the ground and water that has infiltrated. Surface water is captured and carried through watersheds to surface water bodies. Groundwater or phreatic water occupies the pores of the subsoil, giving rise to a zone of saturation, beneath which there must be a layer of impermeable rock.

The water cycle.

A river basin consists of all the land drained by a main watercourse and its tributaries, so that water that reaches the drainage area in the form of precipitation and is not returned to the atmosphere by evaporation and transpiration or does not escape underground to neighbouring basins or to the ocean is eventually filtered, by surface and subsurface routes through the outlet section of the main watercourse of the basin.

Precipitation falling on the basin's land is the source of renewal of its water resources. At the universal level, precipitation phenomena occur at all times. On the other hand, it is also observed that the amount of precipitation varies considerably over time and from place to place, so that certain basins are less favoured than others.

The water resources of all river basins are part of a gigantic circulatory system known as the "hydrological cycle". Although this cycle has no beginning and no end, it is commonly assumed that it starts at the surface of the oceans.

By undergoing continuous evaporation, the oceans provide vapour to the atmosphere, some of which is carried into large air masses and is eventually precipitated in the form of rain, hail or snow, or then condensed in the form of ice or frost on land areas.

This rainwater begins to form part of the catchment resources. The moisture evaporates directly or is then used by vegetation and then returned to the atmosphere. Some of it falls directly into the watercourse beds of the catchment, while some of it reaches the land surface.

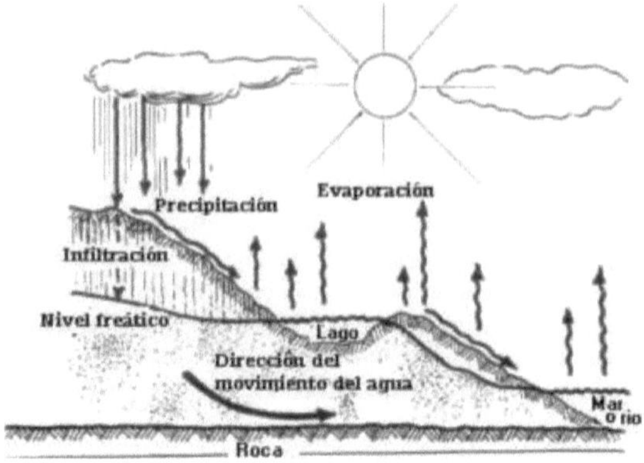

Figure 1.1 The water cycle.

As it reaches the soil surface, only part of the rainfall infiltrates and another part will depend on the difference between the rainfall intensity and the infiltration capacity of the soil.
Of all the water that infiltrates, a part is retained by molecular attraction forces in the so-called soil layer of the aeration zone, to satisfy the moisture deficiency, in relation to the field capacity, produced during the dry period by the phenomena of evaporation and transpiration; if the infiltrated water were sufficient, another part would drain downwards, which could lead to an increase in the underground water storage. In the saturation zone (groundwater zone, where the interstices of the rocks are fully occupied by water) water moves slowly by the action of gravity, in the direction of natural or artificial discharge points and areas. This discharge may occur after a few days, months or even longer periods.
Natural discharge from aquifers can occur by half-springs and by infiltration along river beds; or it can occur by evaporation and transpiration, where the water table is very close to the surface; some of the water is discharged underground into the ocean.
Precipitation, percolation, runoff and evaporation are stages in the water cycle (Fig. 1.1).
Of the water that reaches the earth, some falls directly onto the land, some falls directly onto water surfaces: some flows over the land and makes its way through streams and rivers, ponds, lakes and reservoirs or seas and oceans; some of it returns immediately to the atmosphere by evaporation and transpiration of vegetation and some of it sinks into the ground.
Part of the water that penetrates the earth's crust is retained near the surface, from where some of it evaporates directly, and some is taken up by vegetation, to be returned to the atmosphere by transpiration. The remainder of the infiltrated water drains downwards by gravity, until it reaches the water table, to join the underground reservoir within the earth's crust. Most of the groundwater is discharged to the ground surface through springs and upwelling transmogrifies, or passes either at or below the water table to streams or standing bodies of water, including the oceans.
The water flowing in streams and rivers is derived, only to a small extent, from direct precipitation and mostly from rainwater running off the ground surface, and in more uniform amounts from dry weather flow from the lowering of lake, pond and reservoir levels, as well as from groundwater transmigration.
Evaporation and precipitation are the main driving forces in the water cycle. Solar radiation is the required energy source. Runoff and percolation move the stage for evaporation along the earth's surface; atmospheric circulation for condensation and precipitation.

Natural water sources.

Common sources of freshwater are:

1 Rainwater. Rainwater absorbs gases and vapours normally found in the atmosphere. Rain contains dust, dissolved gases and other substances originating from human activities. Natural pollutants can originate from volcanic activity and wind erosion of the soil. They can be found as

vapour, as a liquid suspended in clouds or falling as rain, hail, snow. It is practically pure, characterised by its lack of mineral salts, it is soft, saturated with oxygen, with a high CO_2 content and therefore corrosive.

2 Spring waters (Fig. 1.2) from rivers and lakes, not yet polluted by man, contain between 0.01% and 0.2% of solid inorganic substances and also dissolved salts, mainly calcium and magnesium compounds, but also, to a lesser extent, sodium, potassium, iron and manganese. The most frequent anions are carbonates, chlorides and sulphates.

Rivers can carry floating organic substances originating from plants and animals. These waters may indeed be drinkable, as long as they do not contain pollutants of industrial origin. Depending on the quality of the soil and subsoil beneath rivers, lakes and on their banks, water can seep into the groundwater at different levels. In these cases, the soil and subsoil material acts as a filter for the water. There are springs and wells of so-called natural mineral waters. Mineral waters contain dissolved salts in high concentrations and dissolved CO_2. In some cases, these waters are thermal and are almost always used for curative purposes.

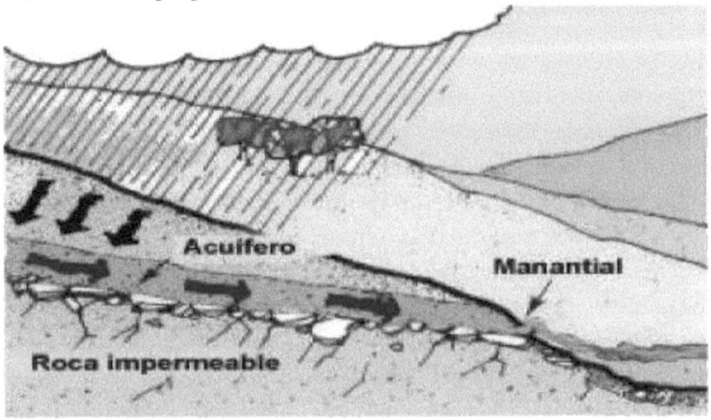

Fig. 1.2 Springs are sources of filtered water.

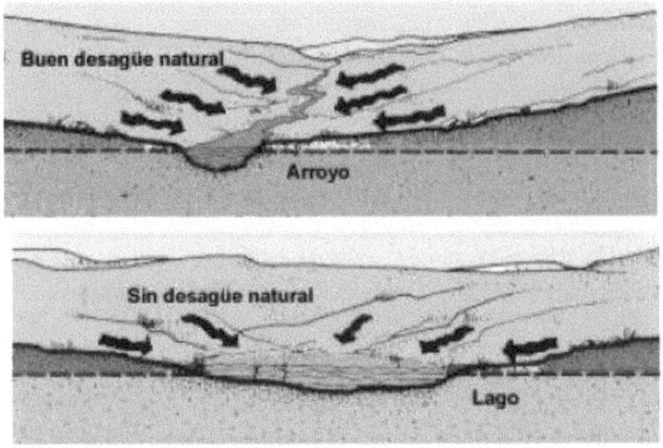

Fig. 1.3 Surface water of natural streams and ponds

3 Surface waters (Fig. 1.3) are those of natural currents such as rivers and streams; and in relative rest in lakes, reservoirs, seas; and in a solid state in ice and snow where they accumulate in large quantities. As they flow over the surface, natural streams are subject to contamination by man and his activities, making them in many cases harmful or unhealthy. Their quality also depends on the type of soil and vegetation.

4 Groundwater, in most cases, originates from rainfall or seepage from rivers and lakes. Groundwater accumulates in porous sediments above impermeable subsoil layers. Their natural groundwater table depends on geological conditions, rainfall and climatic conditions. Human activities

such as mining, river regulation and water abstraction influence the groundwater level. Groundwater usually also flows following the force of gravity. Their velocity depends on geological conditions. When there are cracks and caves in the subsoil, typical for limestone sites, the velocity of their flow increases. Groundwater (Fig. 1.4) is an important source of drinking water.

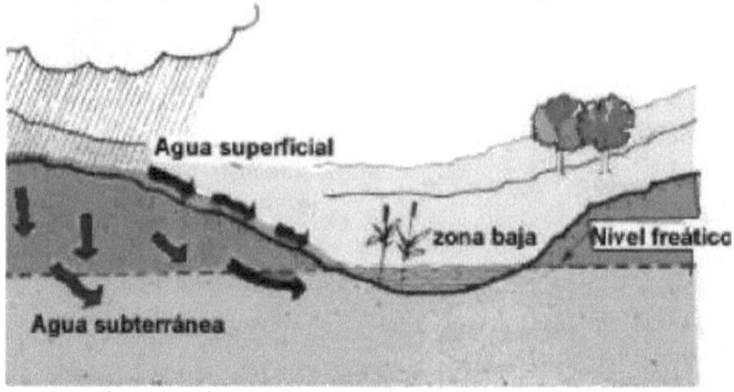

Fig. 1.4 Surface water, groundwater and groundwater table.

These waters are free of microbes when the soil and sediment layers, through which they were filtered, show a sufficient filtration effect and when their permanence in the subsoil is 50 to 60 days.

This is the water that penetrates through the porosities of the soil by means of the filtration process. There are two types of this water: groundwater and artesian water. Groundwater is the water that is contained between the surface of the earth and the first layer or impermeable stratum; it is found in a permeable bed where it moves freely and at atmospheric pressure; it is made up of two zones, a superficial one called the vadose water zone and another that continues up to the impermeable stratum, which is called the saturation zone. The artesian water is the water that is contained between two impermeable strata, it does not move and has a different pressure to the atmospheric pressure. This subterranean water can surface to form springs or feed watercourses or lakes. As it flows through the different layers of the ground, it comes into contact with organic and inorganic substances, some of which are very soluble. By the decomposition of organic matter nitrates or nitrites. They have a low content of dissolved oxygen and a high content of CO_2, which dissolves iron and manganese, substances that are characteristic of these waters. From a bacteriological point of view, they are harmless if they have not been recharged with polluted water.

5 Seawater contains almost 3.5% dissolved salts: the Atlantic Ocean, 3.49%; the Pacific Ocean, 3.46%; and the Arctic Ocean, 3.1%. On average 3% of these salts are sodium chloride (NaCl) and the rest are salts of about 50 different ions. For all living things on land, plants and animals and humans, this water is "poisoned". In order to be drinkable, a special desalination system is needed.

Brackish water on coasts near river mouths, although having a dissolved salt content between 0.1% and 2.5%, is also not suitable as a source of drinking water without a desalination process.

CHAPTER 2

Water and its interaction with solid materials.

There are solid substances that are easily soluble in water and others that are hardly soluble or even insoluble in water. These properties are determined on the one hand by the solid substance, due to its physical and chemical properties, and on the other hand by the properties of water. Due to these solubility properties of the different solid substances in nature. The water cycle has transported many salts from the continents to the oceans, forming salt water, while in most of the springs, streams, rivers and lakes on the continents the water is "fresh" although it also contains dissolved salts, but in low concentrations.

Naturally occurring solids that we commonly consider insoluble, such as rocks, are not insoluble at all. The reason for this is to be found in the chemical equilibrium between the solid substance in contact with water. Three cases are analysed: 1.- Dissolution of CaSO4. A measure to evaluate the solubility of a substance in water is what is known as the solubility product. This is the product of the ion concentrations of a substance in its saturated solution. In the case of gypsum (CaSO4), water does not participate chemically in the equilibrium, but acts only as a solvent. The equilibrium reaction is presented as:

$$CaSO_{4(sólido)} \leftrightarrow Ca^{+2}_{(disuelto)} + SO_4^{-2}{}_{(disuelto)}$$

In this reaction, two phases are present: CaSO4 in the solid phase and calcium and sulphate ions in the liquid phase. If we apply the law of mass action to this equilibrium, we can include the constant concentration of the solid in the equilibrium constant and we have:

$$Kps = [Ca^{+2}]\,[SO_4^{-2}]$$

The Kps constant is called the solubility product, and these solubility products are already determined for many substances. As can be seen in the dissociation reaction of calcium sulphate, two ions are formed which have equal concentrations. The units of Kps are mol $/L^{22}$. Then, if the solubility product at a given temperature is known, the square root must be taken to find the concentration of CaSO4 in dissociated form. If the solubility product of calcium sulphate is 6.1 $x10^{-5}$ mol $/L^{22}$. Then, we can calculate the solubility of gypsum in 100 mL of water for example:

$$6.1 \times 10^{-5} = [Ca^{+2}]\,[SO_4^{-2}] = s \times s = s^2$$

$$s = \sqrt{6.1 \times 10^{-5}} = 7.8 \times 10^{-3} mol/L$$

This is the concentration of CaSO4 in a saturated solution. To convert from moles to grams, multiply by the formula weight of CaSO4 (136.08)(7.8 X 10^{-3}) = 1.06 g per litre.

1.06 x 0.1 = 0.106 g / 100 mL of water

2.- Dissolution of calcium and magnesium carbonate. For this dissolution of carbonate minerals, which most rocks contain, the equilibrium reaction:

$$CaCO_{3(sólido)} \leftrightarrow Ca^{+2}_{(disuelto)} + CO_3^{-2}{}_{(disuelto)}$$

Carbonate anions are formed here, and these anions are part of a series of other equilibria:

$$CO_3^{-2} + H_3O^+ \leftrightarrow HCO_3^{-1} + H_2O$$

$$CO_2 + H_2O \leftrightarrow H_2CO_3$$

$$H_2CO_3 + H_2O \leftrightarrow H_3O^+ + HCO_3^{-1}$$

From this series of balances, it can be seen that both the pH and the presence of carbon dioxide influence the solubility of a carbonate. Calcium and magnesium cations cause the so-called hardness of water.

3.- Dissolution of silicates and SiO_2. Silicon dioxide is found in nature in most rocks as quartz, but there is also an amorphous modification. Both modifications can react with water according to the following reaction:

$$SiO_{2(sólido)} + 2\,H_2O \leftrightarrow Si(OH)_4$$

The dissociation constants of this reaction at a temperature of 25°C are: Kps = 1.99 x 10^{-4} and 1.99 x 10^{-5} respectively for quartz and amorphous quartz. This means that the amorphous species is 10 times more soluble than crystalline quartz.
The ortho silicic acid $Si(OH)_4$ is a very weak acid:

$$Si(OH)_4 + H_2O \leftrightarrow Si(OH)_3O^{-1} + H_3O^{+} \quad Kp = 3.16 \times 10^{-16}$$

$$Si(OH)_3O^{-1} + H_2O \leftrightarrow Si(OH)_2O^{-2} + H_3O^{+} \quad Kp = 2.5 \times 10^{-13}$$

As can be seen from these equilibria, the solubility of quartz will increase under alkaline conditions. The dissolution process of feldspars is more complex. This class of minerals are silicates in which a quarter of the silicon atoms are replaced by aluminium atoms and the resulting negative charges are neutralised by cations of elements of the first or second group of the periodic table. One such feldspar is albite, a sodium aluminium silicate: $NaAlSi_3O_8$.
For feldspars, the chemistry of dissolution is very complex because the reactions are very slow and do not reach complete equilibrium. The reactions usually involve the formation of stable metastable compounds. Also as a result of the reaction with water, new solid phases can be formed. Albite is one such example, where the product is kaolinite, a clay mineral. All feldspars are silicates with a three-dimensional structure, but when altered they produce solution compounds and clay minerals such as kaolinite, which are sheet silicates. In this reaction, hydrogen ions attack the ions in the feldspar structure and some of the released ions are incorporated into a developing clay mineral, while others go into solution.

$$2\,NaAlSi_3O + 2\,H_3O^{+} + 7\,H_2O \leftrightarrow 2\,Na^{+} + 4\,Si(OH)_4 + Al_2Si_2O_5(OH)_4$$

It is noted that acidic conditions are necessary for the reaction to occur. It has been shown that CO_2 in air, dissolved in water (Fig. 2.1) forms hydronium ions (H^+ or $H3O^+$). Therefore, the reaction will be favoured in the presence of CO_2:

$$2\,NaAlSi_3O_{5(sólido)} + 2\,CO_{2(gas)} + 11\,H_2O \leftrightarrow 2\,Na^{+} + 2\,HCO_3^{-1} + 4\,Si(OH)_4 + Al_2Si_2O_5(OH)_{4(sólido)}$$

These examples show how in nature water acts as both a solvent and a reagent, so that over the years it participates in the degradation of rocks, causing the dissolution of minerals that will be an essential part of natural waters. It is very clear that the quality of natural waters will depend directly on the water-rock contact and therefore on the geology of the area.

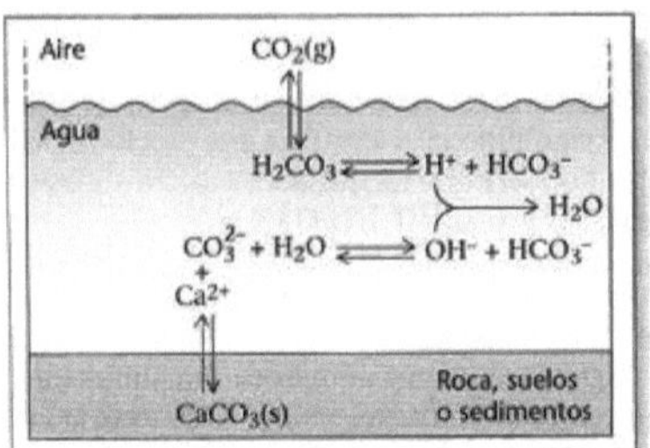

Fig. 2.1 The carbonate system, consisting of its conjugate acids and bases; gaseous carbon dioxide, carbon dioxide, bicarbonate and carbonate ions, together constitute the most important pH regulator in ocean water and groundwater.

CHAPTER 3

Quality and chemical aspects of natural waters.

Natural waters: oceans, rivers, streams, lakes, springs, groundwater and even precipitation in the form of rain, snowfall, hail and fog are never chemically pure. Pure water is "distilled" or "bi-distilled" water. Water in its natural underground pathways and on the earth's surface, as well as precipitation in the atmosphere, always comes into contact with substances that are more or less soluble in water or that undergo reactions with water molecules. The degradation of rocks, the nature and composition of soils and subsoil are decisive for the composition of natural waters with respect to their quality and concentration of dissolved and suspended components. The ions Ca^{+2} and Mg^{+2} result from the interaction between water and carbonate rocks of sedimentary origin. On the other hand, higher concentrations of the silicic acid H_4SiO_4 and lower amounts of calcium and magnesium are found in waters in contact with crystalline rocks. Na^{+} and K^{+} originate from the slow decomposition of feldspars and clays.

Apart from these components, other types of substances may also be found in water. This depends on the geological and geochemical conditions along the underground and surface water course. Sulphate (SO_4^{-2}) comes from gypsum ($CaSO_4$) and iron ores such as pyrite (FeS_2), as well as other sulphides. Phosphate (PO_4^{-3}) and fluoride (F^{-1}) are products of the degradation of apatite [$(PO_4)_3(F,Cl)Ca_5$].

There are springs which contain an excess of inorganic components, trace elements, which are very rare in natural waters. Also volcanic activities with their acidic exhalations have been intensively involved over the years in the transformation of rocks to more soluble minerals. Anions such as carbonate (CO_3^{-2}) and bicarbonate (HCO_3^{-1}) are also present, resulting from the interaction between water and rocks of sedimentary origin, as well as the presence of CO_2 in the air.

All alkaline-earth metal cations are not found on their own, but have their counterparts in the form of anions. Therefore, the concentrations of the cations are inseparably coupled with a complex set of equilibria of dissolution and chemical dissociation of the system:

$$CO_2 \leftrightarrow H_2CO_3 \leftrightarrow HCO_3^{-1} \leftrightarrow CO_3^{-2}$$

The chemical constituents of natural waters.

We need to start with rain, where all natural water comes from. Rain carries with it carbon dioxide (CO_2) which, in combination with water, gives carbon dioxide (H_2CO_3). In turn, the water takes up more carbon dioxide from the soil and humic acid. Therefore, new water can dissolve certain soil materials. The amount of dissolved matter will depend on the chemical characteristics of the soil and subsoil, the time spent in the ground and the distance the water travels in the rocks. As the water remains in the ground it can be concentrated by evaporation and complex changes will occur.

In groundwater we find sodium, potassium, calcium, magnesium, iron, chlorides, sulphates... Silicate may also be present. Fluoride, nitrate and boron, when present, are found in very low concentrations. Carbon dioxide does not always form bicarbonate, but may remain in small portions as a dissolved gas.

Carbon dioxide

As mentioned above, rain contains dissolved carbon dioxide from the air or soil. In combination with water it forms carbon dioxide which is a very weak acid. However, when it comes to water with little of the other chemical elements, the carbon dioxide works strongly, causing a lot of difficulties in the pipes and corroding them. This water acquires iron, the pipes start to leak at the joints and the boilers corrode. To prevent corrosion of water with low mineralisation but with carbon dioxide, its hardness is increased by adding a little calcium oxide.

Calcium bicarbonate

When carbon dioxide water encounters calcareous rocks, calcium bicarbonate Ca $(HCO_3)_2$ is formed. Water with calcium bicarbonate is hard, it requires more soap to make foam. Calcium bicarbonate only remains in solution when it is in equilibrium with carbon dioxide. When heated, the water loses carbon dioxide and the calcium bicarbonate is deposited. In reefs, as seawater warms up near the beach, carbon dioxide escapes and calcium carbonate precipitates, allowing corals to grow. In the springs, calcium carbonate, or travertine, is also deposited. When the water leaves or reaches the surface, its pressure drops and CO_2 escapes. The calcium carbonate is deposited in places like pipes and boxes where the water has a chance to lose gas due to standing still. This deposited calcium carbonate is what is known as "scale".

Calcium sulphate

Carbon dioxide is not needed to dissolve the calcium sulphate and so the simple gypsum solution gives large amounts of calcium. Calcium sulphate also makes the water hard, but the hardness is permanent, i.e. it tends to remain in solution. Such water also needs a lot of soap and causes deposits (incrustations) in pipes and boilers.

The taste of water with too much calcium sulphate is very bad. Water does not quench thirst. In places where the water is "chalky" it would be better to get another type of hydrating drink.

The origin of the sulphate may be in weathered soil, particularly in an arid area where the sediments were not well drained due to a lack of rainfall. АЫ sulphur-bearing minerals can be oxidised to form SO_4^{-2} . Calcium sulphate can also concentrate on the surface by evaporation from wet soil after rainfall or be deposited by evaporation of capillary water.

Magnesium bicarbonate with sulphates

Hard water usually contains magnesium as well as calcium. To dissolve magnesium carbonate, carbon dioxide, in the form of limestone, is needed. Calcium carbonate is more soluble than calcium carbonate but, nevertheless, dolomite is more difficult to dissolve than limestone. Spring deposits, such as travertine, generally contain little magnesium.

Seawater has much more magnesium than calcium due to the greater solubility of magnesium, particularly in a sodium chloride solution. In part, calcium deficiency in the sea is due to extraction by marine organisms.

In the natural waters of the earth the proportion of calcium is generally higher, often higher than that of magnesium. When it happens that an area contains an equal or higher amount of magnesium than calcium, we suspect that it is a water formerly influenced by the sea or partly mixed with seawater.

Sodium bicarbonate

Igneous rocks, particularly acid feldspars, yield sodium bicarbonate after weathering, but generally, the concentration of sodium bicarbonate is low in groundwater. Water with a moderate amount of sodium bicarbonate is very tasty. It is called "mild". For washing such water needs little soap. It is very bad for agriculture. Only when calcium is present in water with excess sodium, it can be used for irrigation. This water is corrosive to aluminium, forming sodium aluminate.

Sodium chloride

Sodium chloride originates from igneous rocks in small quantities, from volcanic gases and from evaporites. The rocks have little chloride, but the volcanic gases are rich in hydrochloric acid and the evaporites commonly have an excess of sodium chloride.

Chloride concentration in a wet area is generally low, perhaps a few parts per million, due to the high solubility of chloride salts and the good circulation of water on land. Near the sea, the wind brings the salt spray to the land, depositing sodium chloride as crystals, which the rain dissolves and carries to the saturation zone.

Water with a high concentration of sodium chloride is very bad. Its taste is unpleasant. It is possible to drink water with a concentration of up to 1000 ppm, but if you drink too much of this water you will get stomach upsets. The standard recommends that in a municipal system the chloride concentration should not exceed 250 ppm.

NOM-127-SSA1-1994, establishes a maximum of 250 ppm.

Iron

Igneous rocks and many ores yield iron through weathering. Sediments have, locally, iron cement. Ferrous materials are a source of iron. It is present in water as bicarbonate or complex salts, perhaps as part of organic salts. Water with little iron has a bad taste. Water with 1 ppm iron can be tested. Used in cooking, laundry and bathroom cleaning, it leaves stains on clothes, utensils and fixtures. The standard recommends that municipal water should have 0.3 ppm or less iron.

Fluor

Fluorine, a minor but not rare constituent, originates from the solution of igneous rocks. Its solubility in a calcareous solution is low, but when the water has a lot of sodium, fluoride may be present. Hard water does not contain fluoride, but when it contains sodium bicarbonate, the presence of fluoride can be elevated. More than 2 ppm of fluoride mixed with water has a detrimental effect on the teeth of children who drink it for the first 12 years. Their teeth become more or less chaste or chaste-stained. Regulations state that water with more than 1.7 ppm should not be used in the public water supply. However, low fluoride has a desirable appearance on children's teeth. Teeth are strengthened and cavities are prevented.

Nitrate

Years ago, nitrate was considered to originate from the organic solution. Contamination by human and livestock faeces. In addition, little nitrate can be fixed in the soil by plant roots. In some places, fertilisers containing nitrates can reach the underground reservoir. The presence of nitrate to some extent is somewhat suspicious. Bacterial depletion may be a cause. Nitrate in a few parts per million is a sign of pollution. But besides that, drinking water with nitrate is very dangerous for pregnant women, as they give birth to blue children with circulation problems "cyanosis". The standard recommends that nitrate should not exceed 45 ppm.

Boron

Boron in concentrations up to 1 ppm in water can damage plants. Therefore, where water is used for irrigation, this constituent needs to be determined.

Boron originates from minerals such as tourmaline. Deposits in dry lakes also contain boron. Boron is common in volcanic gases. In general, natural water does not contain boron.

Major and minor ions in natural waters

In natural groundwater, most of the dissolved substances are in the ionic state. A few of these ions are almost always present and their sum represents almost the totality of the dissolved ions; these are the fundamental ions and most of the chemical and hydrogeochemical aspects will rest on them.

These fundamental ions are:

Cations	Anions
$Ca+^{2}$	$HCO3^{-1}$
Mg+2	$SO4^{-2}$
Na+1	Cl^{-1}
K+1	

The anions nitrate ($NO3^{-2}$) and carbonate ($CO3^{-2}$) and the cation potassium (K+1) are often considered to be in the group of fundamental ions, although their proportion is generally small. Sometimes the ferrous ion ($Fe+^{2}$) is also included.

Among the gases, carbon dioxide (CO_2) and dissolved oxygen (O_2) must be considered as fundamental, although they are not often analysed in groundwater. Among the dissolved substances that are poorly ionised or in a colloidal state, acids and anions derived from silica (SiO_2) are important.

The remaining ions and dissolved substances are generally found in notably smaller quantities than the above and are called minor ions those ions which are usually found in less than 1 % of the total ionic content and trace elements those ions which, although present, are usually in quantities which are difficult to measure by usual chemical means.

The most important minor ions are, in addition to those already mentioned, $NO_3^{-1}, CO_3^{-2}, K^+, Fe^{+2}, NO_2^{-1}, F^{-1}, NH_4^{+1}$ y Sr^{+2} . They are usually found in concentrations between 0.01 and 10 ppm. Minor ions are usually found in concentrations between 0.0001 and 0.1 ppm:

Anions: $B^{-1}, S^{-2}, PO_4^{-3}, H_2BO_3^{-1}, OH^{-1}, NO_2^{-1}, NO_3^{-1}, I^{-1}$

Cations: $Fe^{+3}, Mn^{+2}, Al^{+3}, H^{+}, NH_4^{+1}$

Metal ions derived from As, Sb, Cr, Pb, Cu, Zn, Ba, V, Hg, U, etc., are sometimes present in measurable quantities, but are generally trace elements. All other possible ions are almost always in amounts less than 0.0001 ppm. Although minor ions and trace elements are not usually determined in routine analyses, except in special circumstances, they are nevertheless of interest, especially in studies of origin and relationships between waters and in mineral prospecting. They are currently of particular interest in certain investigations of the effect of substances dissolved in water on public health. The availability of atomic absorption spectrometers makes their analysis much more accessible, although not without difficulties. All of the above refers to the most frequent cases in natural waters; in polluted waters the circumstances can change considerably. In nature, there are sometimes natural groundwaters with compositions which do not conform to the above, either because they contain very small quantities of some of the fundamental ions, or because an important fraction of the ionic content is represented by one or more of the minor or trace ions. Groundwaters called fresh waters contain at most 1000 or perhaps 2000 ppm of dissolved substances; if the content is higher, e.g. up to 5000 ppm, they are called brackish waters and up to 40 000 saline waters. It is not uncommon to find waters with more than 40 000 ppm of dissolved substances, sometimes up to 300 000 ppm. These waters are called brine waters and are often associated with saline deposits, oilfield waters or very old waters at great depths. These waters may contain high amounts of minor or even trace elements and constitute true minerals of these substances, but often remain a small portion of the total content, which is generally dominated by the ions Cl^{-1} and Na^{+1} and sometimes Ca^{+2}.

CHAPTER 4

Determination of physical and chemical parameters in natural waters.

In the study of surface and groundwater, physical and chemical parameters provide the most important information for its characterisation. The following chapters of this text describe in detail some of the physical and chemical measurements and techniques commonly and constantly used in water assessment that can be applied by non-professional hydrochemists, students of chemistry and hydrochemistry or by technicians and/or researchers in the field of water resources, as well as by explorers and anyone interested in the study of water. The techniques presented here, for the characterisation of natural, surface and groundwater, have a micro-scale technical character, where the main objective is to present volumetric analytical techniques that are more accessible and easy to perform, without requiring complex laboratory equipment and utensils, which make the realisation very expensive, as well as the use of the minimum reagents required. They are intended to be field-implemented techniques that are easy to perform at lower cost and in less time. The volumetric techniques presented will also offer a significant reduction in toxic waste disposal, as relatively small quantities will be used.

Sampling

Any water analysis starts with sampling. In order to ensure a high informative value of the analysis results with regard to the quality of the analysed water, it is very important that the samples are representative, especially when it comes to assessing the quality of natural water bodies. It is very important to choose a suitable sampling location to ensure that the sample is representative. The sampling site should not only ensure easy repeat sampling, but also be characterised and marked and preferably easily accessible.

Before each sampling, it is necessary to think about what you want or need to know about the quality of a water. It is very important to establish a sampling plan, before sampling is carried out, in order to prepare and carry all the equipment minimally needed for collection and, if necessary, on-site analysis and, if the sample is taken to the laboratory, to ensure that the quality of the sample does not change during the time between sampling and analysis.

For sampling, it is necessary to know which parameters are to be determined, in order to have the necessary equipment for their evaluation, as some can be carried out directly in the field (e.g. temperature, pH, total dissolved solids, electrical conductivity, ORP). However, in general, for the collection of water samples to be evaluated both in the field and in the laboratory, it is necessary to have perfectly washed bottles and flasks. Each bottle is labelled with an identification label. If this is not possible, at least a few drops of chloroform can be added to each sample to prevent bacteria from reproducing and altering the water, and if they are to be transported to a distant location for analysis, they should be placed on ice (4°C). Samples should never be stored for several days, as the chemical composition may change. They can be kept refrigerated for a short time before analysis.

Sampling for physico-chemical characterisation is easily accessible from the shore, such as in ponds and shallow depressions or large bodies of water. In these cases, the sample is taken directly, for which polyethylene bottles are very suitable, as they are lighter than glass bottles and unbreakable. Water samples should be stored in perfectly clean and closed containers so as not to allow loss of components due to volatility or entry of air (which may favour chemical or bilogical reactions).

For most analyses, refrigeration at a low temperature (4°C) is sufficient to preserve the water samples for a sufficiently long time. At this temperature, the effects of chemical reactions undesirable for the analysis are reduced and the biological activity minimises the chances of decomposition.

Once the samples have been collected, they should be transferred as soon as possible to the analytical laboratory. If the sample components are unstable, portable coolers filled with sufficient ice or any other refrigeration system should be used during transport.

Table 4.1 presents some storage conditions for some of the analytes discussed in this text.

Analyte	Container	Additive	Time max
Alkalinity	Polyethylene		2 weeks
Calcium	Polyethylene	HNO_3 pH< 2	24 weeks
Hardness	Polyethylene	HNO_3 pH< 2	24 weeks
Chlorides	Polyethylene		4 weeks

Table 4.1 Storage conditions for samples, when analyses are not carried out immediately.

Temperature

Heat potential referred to a certain origin, e.g. the melting temperature of ice. Units: Measured in degrees centigrade or Celsius (°C) and in Anglo-Saxon countries in degrees Fahrenheit (°F). °C = 5/9 (°F - 32).

Groundwater has a very low temperature variability, and responds to the annual average of the local atmospheric temperatures, increased by the product of depth and the geothermal gradient (1 °C every 33 m on average, somewhat higher in tectonic and volcanic zones and somewhat lower in large sedimentary basins). Temperature affects the viscosity of the water, gas absorption capacity, etc. Although water temperature is not a chemical characteristic, it affects some of the geochemical properties (such as solubility of gases and minerals). Measurements should be made directly in the field. The temperature should be measured with a thermometer or thermistor. It should be measured in the field as soon as possible to avoid heating or cooling. It must be ensured that the sample represents the temperature of the water in the aquifer, and not inside a catchment, pipe or tank. The ideal water temperature is between 10 and 19 °C.

Fig. 4.1 Digital thermometers and mercury thermometers.

To measure the temperature, the mercury thermometer or probe thermometer should be immersed directly in the water, taking care to do so in the shade, so that it is not directly exposed to the sun. Keep the thermometer in the water for at least 30 seconds to one minute. It is recommended that the reading be recorded with the thermometer in the water to avoid possible variations. The temperature is recorded in degrees Celsius (°C).

Hydrogen potential (pH)

The water molecule is made up of two hydrogen atoms and one oxygen atom, hence its formula H2O. It is a covalently bonded molecule and because of its somewhat asymmetric and strong polarity, water has unique properties from both a chemical and physical point of view. Physically, it is characterised by the fact that it is a liquid between 0 and 100 °C at atmospheric pressure, with a very high specific heat (1 cal/g °C, which is the average value between 0 and 100 °C), and very high heats of vaporisation (540 cal/g at 1 atm) and freezing (80 cal/g). It has the highest known surface tension and readily wets most natural solids. It has a remarkable dilatometric anomaHa, since at 4 °C it has a maximum density (1 kg/litre), being somewhat less dense at 0 °C and considerably less dense in the solid state. Liquid water is only a very weakly ionic body, with molecular dissociation taking place according to the simplified reaction

$$H_2O \leftrightarrow H^+ + OH^{-1}.$$

At any temperature it is true that:

$$[H^+] [OH^{-1}] = K$$

Where K is a constant as a function of temperature, whose value is 1×10^{-14} at 25 °C. If the water is pure, ionic equilibrium requires that $[H^+] = [OH^-]$. The concentration in hydrogenions $[H^+]$ is a very important figure, but to avoid dealing with very small figures, the pH is used, which is defined as $pH = -\log [H^+]$. For pure water at 25 °C it is pH = 7. Water is a chemically very active substance that is very easy to dissolve and react with other substances, both inorganic and organic. Its ionising power and dielectric constant are very high, and it is the most widely used solvent.

Dissolved substances can alter the dissociation equilibrium of the water, thus altering the pH. The dissolution of NaCl hardly modifies this equilibrium, but the addition of HCl or NaOH modifies it very strongly by the addition of H^+ or OH^{-1}, producing acidic or basic water respectively. Many neutral substances, when dissolved, react with water, destroying this equilibrium, as does calcite ($CaCO_3$), which produces a somewhat basic or alkaline solution, as the reaction $H^+ + CO_3^{-2} \leftrightarrow HCO_3^{-1}$, is established, leaving an excess of OH^{-1}. The existence of very weakly ionised acids or bases tends to keep the pH of water almost constant when attempts are made to change it by adding small amounts of strong acids or bases. These solutions are called buffer solutions and are relatively common in nature due to the presence of dissolved CO_2 and HCO_3 in water^{-1}. This explains why most groundwater has a pH between 6 and 8.5.

The pH of water is a measure of alkalinity and acidity and indicates the concentration of hydrogen ions $[H^+]$ and is measured on a scale from 0 to 14. Measurements of this parameter are considered most accurate when measured in the field and very early in the day as pH can be modified by changes in temperature, leakage or incorporation of gases into the sample. The pH generally describes the composition of the water. A pH of 7 indicates that the water is neutral, lower values indicate acidity and higher values indicate a basic character.

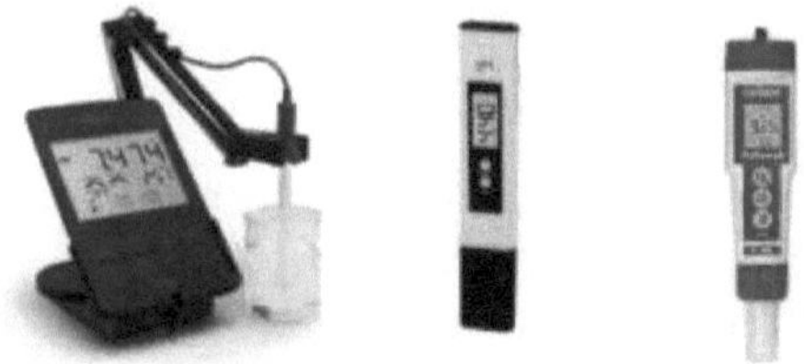

Fig. 4.2 Different models of digital pH meters.

Digital potentiometers (Fig. 4.2), which basically consist of a meter with a glass electrode and a reference electrode, are widely used for pH measurement. For their operation, the use of buffer solutions of pH 4, 6.86 and 9 is required. Adjustment and calibration of the apparatus is carried out according to the procedure indicated in the manufacturer's manual, but in general the following procedure is used:

Rinse the electrode with distilled water. Insert the electrode into the buffer solution and move the power button to the pH position. Move the meter setting to the pH value that corresponds to the buffer solution. Return the button to the off position. Rinse and clean the electrode with distilled water and dry with soft paper.

For the determination of the pH of the water sample:

1) Insert the electrode into the sample and turn the power button to the pH position.
2) Read the pH of the sample, wait for the electrode to reach equilibrium (40 seconds).
3) Return the button to the off position.
4) Rinse the electrode with distilled water and keep it immersed in water while not in use and/or cover the electrode.
5) The pH value is read directly from the potentiometer dial.

pH measurement with indicator paper.

If a pH meter is not available, pH paper can be used, which is a coloured indicator that can be used to give different colours over the full pH range. This method will be an approximation, as it is inaccurate. To do this, a strip of paper is inserted into the water sample. This will produce a series of colours which can be compared visually with a colour scale. The approximate value is noted and the paper strip is destroyed.

Electrical conductivity (EC).

Also known as resistivity. Conductivity is the electrical carrying capacity of a conductor. In the case of water, it is the ability to conduct electricity. The electrical conductivity (EC) in water is characterised by the content of dissolved salts. Completely pure water is not electrically conductive. Conductivity is measured in microsiemens per centimetre (^S/cm) or microohms per centimetre (цO/cm). Conductivity increases with temperature, so it should be measured over a reference temperature (18°C or 25°C). As the electric current is carried by ions in solution, the more ions there are in the water, the higher the conductivity. Therefore, the value of the EC measurement is used as a parameter to estimate the total amount of total dissolved solids (TDS). The total dissolved solids in the water in ppm or mg/L should be equal to the conductivity value of the sample multiplied by a factor between 0.55 and 0.7. The following equation can be used to calculate the amount of total dissolved solids (TDS) from the electrical conductivity (EC):

$$\text{SDT (mg/L)} = \text{CE } (\mu\text{S/cm})\ (550 - 700)$$

Conductivity is determined by comparing the measurement of the electrical resistance between two electrodes of a water sample with that of a standard solution of potassium chloride at 25°C. The electrical conductivity in natural fresh water can vary between 100 and 2000 pS/cm. Nowadays, digital portable meters are widely used (Fig. 4.3).

Fig. 4.3 Electrical conductivity meters, currently in use.

Total Dissolved Solids (TDS)

It is a parameter used to evaluate the content of dissolved ions in water, i.e. its content of dissociated mineral substances. It is recommended to do the test at the natural water sampling site.

Water is a solvent capable of dissolving most solids to some degree. Of the majority of solutes found in groundwater, only a small part is present in an amount greater than one ppm under typical natural conditions. Of the total dissolved solutes in water, major ions occupy 95-99% of the total, minor ions 1 to 3% and trace ions about 0.1%. The solutes that are in higher concentrations than the above are called major ions, and those that are in lower amounts are called trace elements. It is important to note that the relative abundance of the elements that make up the solutes dissolved in water depends on their chemical mobility. As groundwater migrates underground from recharge zones to discharge zones, it changes its chemical composition by dissolving small amounts of materials in its path, increasing the amount of dissolved solids it contains. Official standards preferably set up to 500 ppm, with up to 1000 ppm of total dissolved solids tolerated in potable freshwaters. Digital meters Fig. 4.4 are currently used for their determination.

Redox potential (ORP)

The redox potential measures the tendency of a system to oxidise or reduce a chemical species. Redox systems are in direct relation to pH, dissolved oxygen, pressure and temperature. The redox potential is very important in salt dissolution processes due to the attack of water on minerals, since chemical species can change their valence state.

The ORP or oxidation-reduction potential (ORP) is used as an effective measure of sanitation activity in drinking water, swimming pools and spas.

ORP electrodes were studied at Harvard University in 1936. These studies showed a strong correlation of ORP and bacterial activity. In water disinfection, the concentration of free chlorine and the contact time with the water as well as pH and temperature are very important.

The continuous measurement of the redox potential of the treated water allows the adjustment of the quantity of bio-oxidising agent added to the water (dosage), required to react with organic and inorganic substances, and reduce the concentrations of the bacterial populations contained in the water to the desired levels.

The World Health Organization (WHO), adopted in 1971 a value of 650 mV as an adequate value for drinking water, in general it can be considered that with this value maintained for 30 minutes the water is adequately disinfected, although it will be necessary to carry out an individualised study for each case. The ORP has no direct relation with the concentration in ppm of disinfectant as it measures the oxidation activity in the water and not the concentration of oxidant (chlorine, ozone, and other oxidative disinfectants).

The ORP is measured electrometrically using a reference electrode (Fig. 4.4). The ORP should always be determined at the place where the sample is taken, as small changes in the ambient conductivity can cause significant changes in the ORP value.

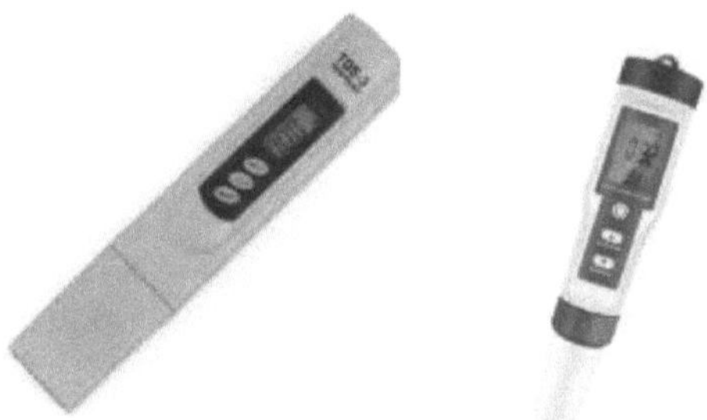

Fig. 4.4 Digital SDT and ORP meters. Multiparametric meters are now available, easy to transport and with reliable results.

CHAPTER 5

Determination of Chlorides.

Chlorides in natural waters.

The chloride ion Cl^- , is mainly associated with rainfall and its content depends on the height and distance of the sampled point from the sea, as well as on the type of rainfall event. In wells located on the coastal plains, a high Cl content$^-$ may be associated with saline intrusion or inputs from polluting sources. This ion predominates in seawater and in many freshwaters, particularly those associated with marine shales and evaporites. It generally forms very soluble salts. It is very stable in solution. It is not oxidised or reduced in natural waters. It is generally associated with the Na ion$^+$, especially in very saline waters, but the number of milliequivalents of the two ions is not necessarily equal. Concentrations of this ion are between 10 and 250 ppm in fresh waters, but it is not uncommon to find much higher contents up to several thousand ppm. More than 300 ppm gives a salty taste to drinking water, but it is not harmful at least up to a few thousand ppm. It is essential for life. High levels are harmful to many plants and communicate corrosiveness to water. The limit allowed by International Standards is 250 ppm for drinking water.

For analysis in the laboratory, titration with AgNO3 is used, using Ag2CrO4 as an indicator (Mohr's method) with a yellow to reddish-orange change.

Mohr's method

In volumetric analysis, there is a group of substitution reactions in which one of the products is insoluble, and therefore the methods which are based on the formation of a precipitate are called precipitation volumetry. The most important reactions of this type involve silver ions [Ag+], which is why it is called Argentometry. Karl Friedrich Mohr, in 1856, proposed the use of a potassium chromate solution as an indicator of the end of the reaction between chlorine ions [Cl^-] and silver ions [Ag^+]. This method, known by the name of its author, is based on the different solubilities of silver chloride (AgCl) and silver chromate (Ag2CrO4) and can be explained as follows: if a small quantity of potassium chromate is added to a neutral solution of a chloride and titrated with a titrated solution of silver nitrate (AgNO3), there is a tendency for two precipitates to form: that of silver chloride (AgCl, white), and that of silver chromate (Ag2CrO4, orange red), but the former being more insoluble, as long as there are chloride ions [Cl^-] in the solution, the formation of silver chloride will take place; and only when all the chloride has precipitated, a slight excess of silver ions [Ag^+] will produce silver chromate, which remains and imparts a reddish-orange colour to the liquid, indicating the end of the titration.

Theoretical principle.

This methodology starts in principle from the equilibrium on which the reaction between [Ag+] ions and [Cl^-] ions is based, according to the reaction:

$$Ag^+ + Cl^- \rightarrow AgCl\ (\text{blanco})\ Kps = 1.8 \times 10^{-10}$$

Titration curves for chloride ion [Cl^-] = 0.1 M, using 0.1 M AgNO3 are presented, as in Fig. 5.1.

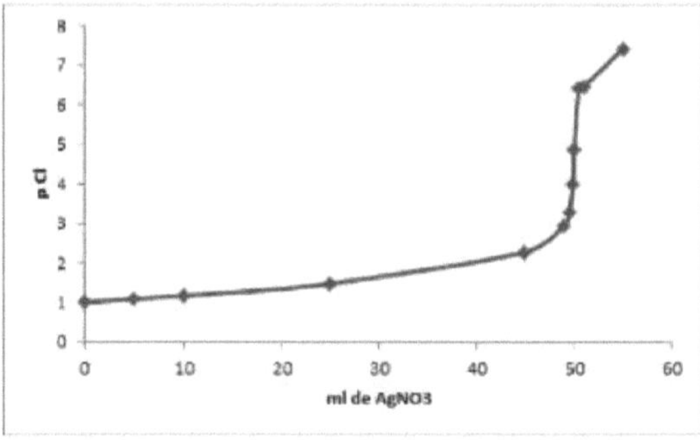

Fig. 5.1 Chloride titration curve with silver nitrate.

At the equivalence point, the $[Cl^-] = [Ag^+]$, then:

$$[Cl^-] = [Ag^+] = \sqrt{Kps} \quad \sqrt{}$$

$$\text{Y el pCl} = -\log[1.34 \times 10^{-5}] = 4.87$$

In the determination of chloride ion by titration with silver, the end point should occur when the concentration of the ion being titrated has fallen to 1.34×10^{-5} mol/L or when the concentration of the titrant ion is slightly above 1.34×10^{-5} mol/L.

The end point in this method is determined by the first formation of an orange-red precipitate of silver chromate which will appear when the precipitation of the silver chloride is complete; the indicator will be a solution of potassium chromate (K_2CrO_4).

$$Ag^+ + Cl^- \rightarrow AgCl\ (\text{blanco}),\ Kps = 1.8 \times 10^{-10}$$

$$2\,Ag^+ + CrO_4^{-2} \rightarrow Ag_2CrO_4\ (\text{rojo}),\ Kps = 1.3 \times 10^{-12}$$

As silver chromate is more soluble than silver chloride. When silver ions are added to a solution containing a high concentration of chloride ions and a low concentration of chromate ions, the silver chloride will precipitate first; silver chromate will not form until the concentration of silver increases sufficiently to exceed the Kps of silver chromate.

Knowing that the sedimentation of any precipitate begins after the product of the concentrations of the constituent ions becomes greater than the solubility product, it is not difficult to calculate what the concentration of the [CrO4 ion^{-2}] must be for the Ag2CrO4 precipitate to form only when practically all the Cl ions^{-1} are titrated. That is, it is possible to calculate the concentration of chromate that will cause the precipitation of silver chromate at the equivalence point:

$$Kps\ Ag_2CrO_4 = 1.3 \times 10^{-12} \text{ therefore:}$$

$$[Ag^+]^2\,[CrO_4^{-2}] = 1.3 \times 10^{-12} \text{ whence:}$$

$$[CrO_4^{-2}] = \frac{1.3 \times 10^{-12}}{[Ag+]^2}$$

And as at the equivalence point $[Cl^-] = [Ag^+] = 1.34 \times 10^{-5}$

$$[CrO_4^{-2}] = \frac{1.3 \times 10^{-12}}{(1.34 \times 10^{-5})^2} = 0.0072\ M$$

According to this calculation, the chromate ion concentration must be slightly higher than 0.0072 mol/L for Ag2CrO4 to form at the equivalence point of the titration with silver ion. In practice, a 5% solution of K2CrO4 meets the expectations to be a good indicator for this titration.

It should be noted that, since the Ag2CrO4 precipitate dissolves well in acids, all determinations carried out by this method must be performed in a neutral or very weakly alkaline medium within the rather narrow pH limits, i.e. 7 to 10.

Material: 1 x 1 mL micro burette

1 micro holder

1 volumetric burette of 3 mL

1 x 25 mL Erlenmeyer flask

Reagents:

Potassium chromate: Dissolve 5 g of Ag2CrO4 in 100 mL of distilled water. Store in a dropper bottle.

2.- Titrated solution of AgNO3 0.01 N

Procedure:

1 Take a 3 mL water sample and place it in the conical flask.

2 Add one drop of 5% potassium chromate solution.

3 Proceed to titrate with the 0.01 N AgNO solution$_3$ until a reddish-orange colouration appears.

Calculations:

The concentration of chloride ion in the sample expressed in mg/L can be obtained from the following expression:

Example: A 3 mL sample of water requires 0.23 mL of a 0.01 M AgNO3 solution. Calculate the chloride concentration.

Microscale volumetric methods for water analysis in the field.

Surface and groundwater sampling.

CHAPTER 6

Alkalinity Determination

Alkalinity in natural waters.

The alkalinity of water is its ability to neutralise acids, and is the sum of all titratable bases. The measured value can vary significantly with the endpoint pH used. Alkalinity is a measure of an aggregate property of water and can be interpreted in terms of specific substances only when the chemical composition of the sample is known. Because the alkalinity of many surface waters is primarily a function of carbonate (CO3 =), bicarbonate (HCO3 -) and hydroxide (OH-) content, it is taken as an indicator of the concentration of these constituents. Measured values may also include contributions from borates, phosphates, silicates or other bases that are present. It is a measure of the ability of a water to neutralise a strong acid. In natural waters this capacity can be attributed to the concentrations of HCO_3^-, CO_3^{-2} and OH^-.

The origin of the presence of carbonate and bicarbonate in natural or surface waters can be diverse: dissolution of atmospheric carbon dioxide.

$$CO_2 + 2H_2O \leftrightarrow H_3O^+ + HCO_3^-$$

The dissolution of carbonate rocks:

$$CaCO_3 + 2H_3O^- \leftrightarrow Ca^{+2} + 2HCO_3^- + 2H_2O$$

$$HCO_3^- + H_3O^+ \leftrightarrow H_2CO_3 + H_2O$$

$$H_2CO_3 \leftrightarrow CO_2 + H_2O$$

Therefore, alkalinity is also equal to carbonate hardness and is the quantitative capacity of a water to react with hydronium ions $H3O^+$ or H^+. Generally, alkalinity is the sum of the content of carbonate, bicarbonate and hydroxide ions, but it can also include other anions such as borates, phosphates, silicates. Fortunately, since the pH of natural water is usually between 6 and 8.5, the predominant anion will be the bicarbonate HCO_3^-, as shown in Figure 6.1 .

It is a measure of the neutralising capacity of an acid. It is used to determine the amounts of carbonate and bicarbonate in water, under the "assumption" that other chemical species are considered minimal. Carbonate and bicarbonate impart alkalinity to water, giving it acid-consuming capacity. The dissolution of limestone and dolomite rocks is the source of bicarbonate and carbonate. Fresh water can have between 50 and 350 ppm of bicarbonate (ground water about 200 ppm). Carbonate ion is in much lower concentrations than bicarbonate. If the pH < 8.3 it is considered zero. In alkaline waters with pH > 8.3 significant amounts of carbonate may be present. Normally, all surface and ground freshwaters have a pH < 8.3, so the alkalinity will only be a function of the concentration of the bicarbonate ion, HCO_3^-.

According to the Secretaria de Salubridad y Asistencia, the maximum amount of total alkalinity, expressed as CaCO3, that water for human consumption can contain is up to 400 ppm.

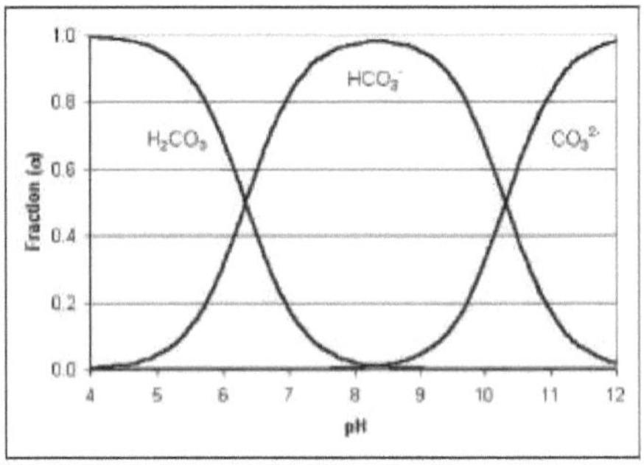

Fig. 6.1 Influence of pH on the equilibrium of carbonate species. In natural freshwaters the pH is between 6 and 9. Therefore, the dominant species will be the bicarbonate HCO_3^-

Theoretical principle

In the determination of alkalinity a known volume of sample is titrated with a standard solution of a strong acid to a pH value in the range of approximately 3.8 to 4.8. This end point is almost always detected by the colour change of the methyl orange indicator. The added H+ is the stoichiometric amount that is required for the following reactions; thus, it is assumed that at a pH of 8.3, the reactions will take place:

$$OH^- + H^+ \leftrightarrow H_2O \text{ y } CO_3^{-2} + H^+ \leftrightarrow HCO_3^-$$

They are complete, and at around a pH of 3.8 to 4.8, the reaction

$$HCO_3^- + H^+ \leftrightarrow H_2CO_3$$

It is complete. If the phenolphthalerna indicator is applied to a water sample and it turns colourless, it is established that the only species causing the alkalinity is the bicarbonate HCO_3^-.

In the course of the interaction between the bicarbonate ion and the strong acid the reactions take place:

$$H_2CO_3 \leftrightarrow H^+ + HCO_3^{-2},\ K_1 = 4.6 \times 10^{-7},\ pK_1 = 6.34$$

$$HCO_3^- \leftrightarrow H^+ + CO_3^{-2},\ K_2 = 5.6 \times 10^{-11},\ pK_2 = 10.25$$

Thus, at the beginning of the titration there is the bicarbonate solution and at the equivalence point the H_2CO_3 solution. The pH of the solution is therefore determined by the ratio of these two substances. Assuming the titration of a 0.01 M bicarbonate solution with a 0.01 M strong acid solution. The pH value of the bicarbonate starting solution can be calculated according to the formula for finding the pH of weak dibasic acid solutions:

$$pH = \frac{pK_1 + pK_2}{2} = \frac{6.34 + 10.25}{2} = 8.3$$

The end point of this titration is determined on the assumption that at the equivalence point all the bicarbonate is transformed into the weak acid H_2CO_3, whose molar concentration will be approximately 0.05 M, therefore:

$$pH = \frac{pK_1 + pC_{ácido}}{2} = \frac{6.34 + 1.30}{2} = 3.84$$

The pH value falls in the mid-point of the range of the methyl orange shift (3.1 to 4.4). This is the basis for the determination of alkalinity in a mixture of hydroxides or carbonates or carbonates and bicarbonates, because when a titrated acid is added in the presence of phenolphthalerine, the colour change takes place when all the hydroxides are neutralised and only half of the carbonates change to bicarbonates. The bicarbonate formed produces a solution whose pH is 8.3, i.e. it discolours the phenolphthalerine; if methyl orange is added, against which the solution is still alkaline, since this indicator changes between 3.2 and 4.4, by adding titrated acid until the yellow colour changes to an orange-mamey colour.

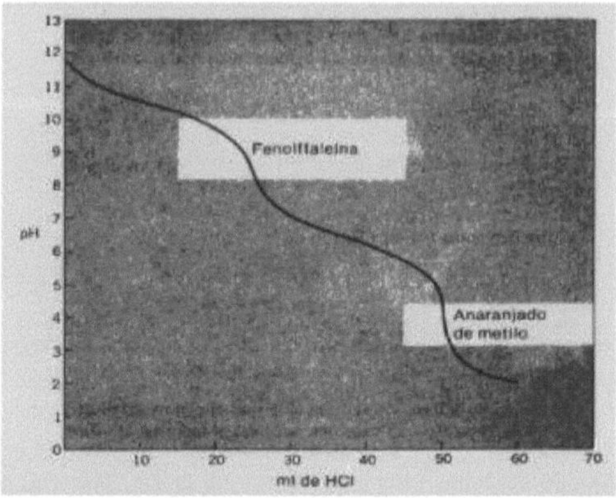

Figure 6.2 Titration curve of a carbonate with strong acid.

What is customary is to titrate the carbonate ion as a base with a strong acid; in this case two inflections are obtained as shown in figure 6.2.

Fortunately, natural waters are found with pH < 8.3 and therefore, it is common that the predominant form of alkalinity is bicarbonates, so it can be assumed that this form generates most of the total alkalinity in the sample.

Material:

1 x 1 mL micro burette

1 micro support

1 x 3 mL volumetric pipette

1 x 25 mL Erlenmeyer flask

Reagents:

1. Methyl orange solution. Dissolve 0.05 g of the indicator in 100 mL of distilled water.
2. Titrated solution of H2SO4 0.02 N

Procedure:

1 Take a sample of 3 mL of water and place it in the 25 mL Erlenmeyer flask.

2 Add 1 or 2 drops of methyl orange indicator.

3 Proceed to titrate the solution with H2SO4 0.02 N, until the colour changes from yellow to mamey.

Calculations:

The total alkalinity, expressed as mg $_{CaCO_3}$, per litre, is calculated by the following expression:

$$CaCO_3 = \frac{V_{ácido} N_{ácido} (50000)}{V_{muestra}}$$

Total alkalinity como

Example: The pH of a water sample is 7.47. A 3 mL sample is titrated with 0.78 mL of a 0.02 N H2SO4 solution.[6] What will be the alkalinity of this water?

$$Alkalinity = \frac{(0.78)(0.02)(50000)}{3} = 260\ ppm$$

To calculate the concentration of [HCO_3^{-1}] the following relationship can be used:

$$[HCO_3^-] = \frac{V_{ácido} N_{ácido} (61\,000)}{V_{muestra}}$$

CHAPTER 7

Hardness Determination.

Hardness in natural waters.

Natural waters contain various quantities of salts and are of very variable composition. Some of them, when used for washing with ordinary soaps, precipitate clusters and are therefore called "hard" waters. The hardness of water is mainly due to the amount of dissolved alkaline earth metals. Most of these ions are the cations Ca^{+2} and Mg^{+2}. The presence of carbonates, sulphates and chlorides of calcium and magnesium cause water hardness. Dissolved divalent and trivalent metal ions such as iron, manganese, aluminium and zinc, although in smaller quantities, also contribute to water hardness. Hard water is found mainly in regions where limestone rocks are abundant. Soft water is found in regions where crystalline rocks such as granite, gneiss, basalt and schist dominate. Soft water contains much less of these alkaline earths and is always better for any application. Rainwater is also soft.

The dissolved alkaline-earth cations are not found alone in water, but have as their counterparts anions mainly carbonate (CO_3^{-2}), bicarbonate (HCO_3^-) and sulphate (SO_4^{-2}). Therefore, the concentrations of the cations depend on a complex set of equilibria of dissolution and chemical dissociation of the system: carbonate^bicarbonate^carbon dioxide.

In the case of calcium and magnesium carbonates such as limestone or dolomite, in addition to the equilibrium:

$$CaCO_{3(sólido)} \leftrightarrow Ca^{+2}_{(disuelto)} + CO_3^{-2}{}_{(disuelto)}$$

In which carbonate anions are formed, these anions are part of another series of equilibria:

$$CO_3^{-2} + H_3O^+ \leftrightarrow HCO_3^- + H_2O$$

$$CO_{2(gas)} + H_2O \leftrightarrow H_2CO_3$$

$$H_2CO_3 + H_2O \leftrightarrow H_3O^+ + HCO_3^-$$

From this series of equilibria, it can be seen that both pH and the presence of carbon dioxide influence the solubility of carbonates.

Calcium is the fifth most abundant element in the earth's crust. It is present in most water due to the high solubility of the rocks that contain it. Calcium salts are responsible for the formation of deposits in pipes and water heating systems.

The origin of magnesium in water is explained by the dissolution of rocks containing ferromagnetic minerals (clays) and some carbonate rocks (dolomite). It is also found in many organometallic compounds and in natural organic matter.

The classification of water with respect to the amount of mg per litre of calcium carbonate is classified as follows:

Hardness as $CaCO_3$ (mg/L)	Type of water
0 a 75	Freshwater
75 a 150	Medium hard water
150 a 300	Hard water
>300	Very hard water

Theoretical principle for the determination of total hardness

The method for determining the hardness of water involves the use of solutions of ethylenediaminetetraacetic acid or its sodium salts as a titrating agent. These solutions form "soluble complex ions" with calcium, magnesium and other hardness-causing ions. In the 1940s Schwartzenbach introduced an important group of reagents that form chelate complexes with metals. These reagents are known as "complexones", the most important of which is $(HOOCCH_2)_2N\text{-}CH_2\text{-}CH_2\text{-}N(CH_2COOH)_2$, usually called EDTA and represented as H_4Y. It is insoluble in water, but its disodium salt

Na2H2Y2 is very soluble:

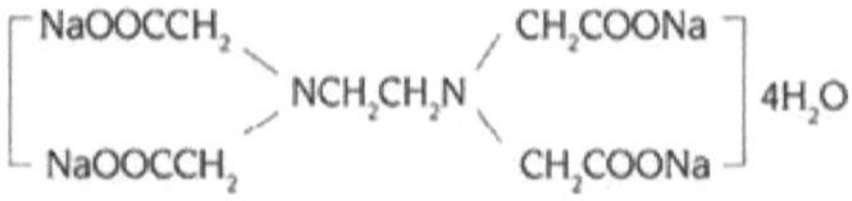

The neutral species of EDTA corresponds to a tetraprotic acid (H4Y). Some of its dissociation equilibria are presented below:

$$H_4Y + H_2O \leftrightarrow H_3O^+ + H_3Y^- \quad Ka1 = 1.02 \times 10^{-2}$$

$$H_3Y^- + H_2O \leftrightarrow H_3O^+ + H_2Y^{-2} \quad Ka2 = 2.14 \times 10^{-3}$$

$$H_2Y^{-2} + H_2O \leftrightarrow H_3O^+ + HY^{-3} \quad Ka3 = 6.92 \times 10^{-7}$$

$$HY^{-3} + H_2O \leftrightarrow H_3O^+ + Y^{-4} \quad Ka4 = 5.50 \times 10^{-11}$$

As in all polyprotic acids, the abundance of the different species depends on the pH of the medium. This can be seen in the figure, where the fractions of each species are plotted as a function of the pH. In figure 7.1, the distribution of the EDTA species as a function of pH is shown.

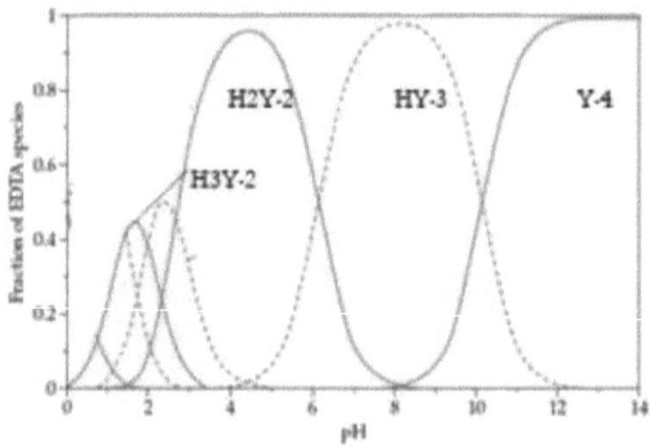

Fig. 7.1 Distribution of EDTA species as a function of pH.

EDTA forms 1:1 complexes with the metal ions, which can be schematized as follows:

$$M^+ + H_2Y^{-2} \leftrightarrow MY^{-3} + 2H^+$$

$$M^{+2} + H_2Y^{-2} \leftrightarrow MY^{-2} + 2H^+$$

$$M^{+3} + H_2Y^{-2} \leftrightarrow MY^{-1} + 2H^+$$

Cations with a high ionic charge form the most stable complexes and exist in more acidic solutions. The three-dimensional structure of the metal-EDTA complexes is shown in Figure 7.2.

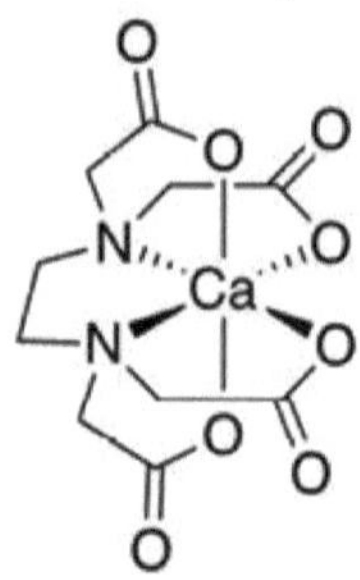

Fig. 7.2 Three-dimensional structure of the Ca-EDTA complex

Calcium, like magnesium, forms very similar structures which can be expressed as follows:

$$Ca^{+2} + H_2Y^{-2} \rightarrow CaY^{-2} + 2\,H^{+}$$

$$Mg^{+2} + H_2Y^{-2} \rightarrow MgY^{-2} + 2\,H^{+}$$

The indicator for the reaction, when Ca+2 and Mg+2 are titrated in water, should reveal the presence of a slight excess of Na2Y-2.

A widely used indicator for this purpose is erythiochrome black T, which is used to indicate when all calcium and magnesium ions have formed a complex with EDTA. When a small amount of blue eriochrome black T is added to water at pH = 10, the indicator combines with Mg+2 according to the following reaction:

$$Mg^{+2} + HIn^{+2}\,(azul) \rightarrow MgIn^{-1}\,(rojo) + H^{+}$$

During titration of a solution containing both Ca^{+2} and Mg^{+2}, EDTA will react as follows:

1. First with Ca^{+2} until the calcium concentration is very low:

$$Ca^{+2} + H_2Y^{-2} \rightarrow CaY^{-2} + 2H^{+}$$

2. Then with the Mg^{+2}

$$Mg^{+2} + H_2Y^{-2} \rightarrow MgY^{-2} + 2H^{+}$$

3. Then with the red-coloured magnesium-based indicator complex, $MgIn^{-1}$ or Mg-NET. The disappearance of part of the red colour identifies the turning point. This disappearance is caused by the slight excess of EDTA, which separates the red Mg^{+2} from the red $Mg\text{-}In^{-1}$, leaving the blue HIn^{-2} (stable at pH=10).

$$\text{Mg-NET (rojo)} + \text{EDTA} \leftrightarrow \text{Mg-EDTA} + \text{NET (azul), o}$$

$$MgIn^{-1}\,(rojo) + H_2Y^{-2} \leftrightarrow MgY^{-2} + HIn^{-2}\,(azul) + H^{+}$$

Calcium hardness

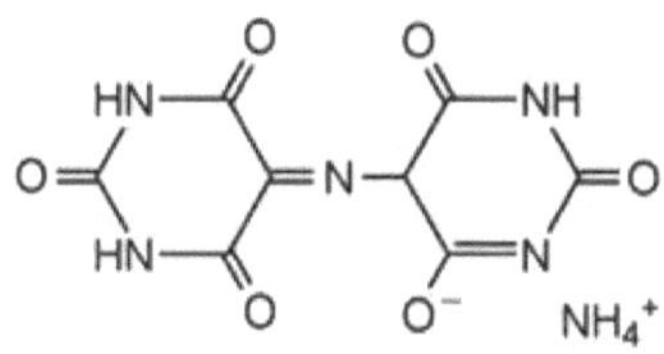

When EDTA solution is added to water containing calcium and magnesium, the EDTA reacts with the calcium and magnesium.

first with calcium and then with magnesium. If the pH is raised sufficiently so that the magnesium precipitates as hydroxide and an indicator is used, which is combined only with calcium, the calcium content of the water can be determined directly. The indicator used is ammonium purpurate or murexide, whose formula is: C8H4O6N5 (NH4), mixed with sodium chloride. The reaction can take the following form: The Ca ion^{+2}, reacts with the purple murexide sodium salt to give a pink complex:

$$2\,(C_8H_4O_6N_5Na)\,(\text{púrpura}) + Ca^{+2} \leftrightarrow (C_8H_4O_6N_5)_2Ca\,(\text{rosa})$$

(pink)

When EDTA is titrated with EDTA, it forms a complex first with the calcium that was in the solution and then with the calcium that has been incorporated into the indicator, returning it to its original colour and indicating the end point of the titration. The reaction may be indicated as follows:

$$[Mur]_2Ca\,(\text{rosa}) + \text{EDTA} \leftrightarrow \text{EDTA-Ca} + 2\,[Mur]\,(\text{purpura})$$

(purple)

The EDTA titration should be done at alkaline pH (between 12 and 13), so that the magnesium ion is precipitated as hydroxide and does not react with the EDTA.

Total hardness (Ca + Mg)

The water sample is brought to pH 10 using ammonium chloride-ammonium hydroxide buffer and adding eriochrome-T black as an indicator. The indicator forms a weakly dissociated, wine-red compound with the magnesium ions. During titration with EDTA titration solution, first the calcium ions are removed from the solution, then the magnesium ions and, at the end point, the magnesium, from the wine-red, magnesium- eriochrome-T-black complex producing a distinct colour change from wine-red to blue.

Total hardness in fresh water can reach a maximum concentration of up to 300 mg/L.

Material:

1 x 1 mL micro burette

1 micro holder

1 volumetric pipette of 3 mL

1 x 25 mL Erlenmeyer flask

Reagents:

EDTA 0.01 M

Eriochrome-T black

Buffer solution at pH = 10 (Dissolve 16.9 g of ammonium chloride in 143 mL of concentrated ammonium hydroxide and dilute to 260 mL in distilled water).

Procedure:

1 Place 3 mL of the water sample in the Erlenmeyer flask.

2 Add 2 drops of the Buffer solution at pH = 10 and shake the sample.

3 Using a capillary tube or a small spatula, place a few crystallites of the black eriochrome-T indicator (the solution will take on a wine-red colour).

4 Titrate with the 0.01 M EDTA solution, drop by drop and shaking until the solution changes from wine red to blue.

The water hardness, expressed in mg $_{CaCO3/L}$, shall be calculated by means of the equation:

$$Dureza = \frac{V_{EDTA} M_{EDTA} (100)(1000)}{V_{MUESTRA}}$$

Calcium hardness:

The water sample is brought to a pH of 12 using sodium hydroxide solution for the precipitation of magnesium. The indicator murexide forms a pink complex with the calcium. It is then titrated with EDTA solution to the equivalence point and the solution turns a purple colour.

Material:

1 x 1 mL micro burette

1 micro holder

1 3 mL volumetric pipette

1 x 25 mL Erlenmeyer flask

Reagents:

EDTA 0.01 M

Sodium Hydroxide 1 M

Murexide 1% in sodium chloride

Procedure:

1 Place 3 mL of the water sample in the conical flask.

2 Add 2 drops of 1M sodium hydroxide solution to achieve a pH of approximately 12.

3 With the aid of a small spatula, add a very small quantity of the murexide indicator and stir thoroughly until a pink colouring is obtained.

4 Titrate with 0.01 M EDTA until the solution turns from pale pink to purple. To better appreciate this change, it is recommended to prepare a blank, using distilled water, sodium hydroxide solution and a pinch of murexide indicator.

The concentration of calcium ion, expressed in mg/L, shall be calculated by means of the following equation:

$$Calcio = \frac{V_{EDTA}M_{EDTA}(40.08)(1000)}{V_{MUESTRA}}$$

The magnesium hardness can be calculated as the difference between the total hardness and the calcium content. The magnesium concentration, expressed in mg/L, is determined by the difference between the two (taking into account the molecular weights):

$$Magnesio = \left[Dureza - \frac{Calcio\ (100)}{40.08}\right]\frac{24.30}{100}$$

Example: A 3 mL sample of spring water is conditioned to pH=10 with buffer solution and a micro pinch of eriochrome black T is added. It is titrated with 0.74 mL of 0.01M EDTA solution. Calculate the water hardness.

$$Dureza = \frac{(0.74)(0.01)(100000)}{3} = 246.66\ ppm$$

Example: A 3 mL sample of spring water is brought to pH= 12 with NaOH and a micro pinch of murexide indicator is added. Titrate with 0.58 mL of 0.01 M EDTA solution. Calculate the concentration of Calcium and Magnesium in the water sample.

Example: A 3 mL water sample is placed in a 25 mL flask and 2 drops of buffer, pH=10, are added. A micro pinch of eriochrome black T is added. It is then titrated with a 0.01 M EDTA solution, using 0.87 mL. Another sample of 3 mL of water is brought to pH 12 with NaOH and a micro pinch of murexide indicator is added. Titrate with 0.73 mL of 0.01 M EDTA. Calculate the total hardness and the concentrations of calcium and magnesium.

[290

Stalactite drip water sampling inside the cave "El Nuevo San Joaqum" in the Sierra Gorda Queretana.

CHAPTER 8

Determination of sulphates.

Sulphates in natural waters.

Sulphates (SO_4^{-2}), after bicarbonates, are the main anions present in water; they can occur naturally or as a consequence of industrial water discharges and the use of agricultural fertilisers. When sulphates occur naturally, it is possible that their origin is due to natural mineral deposits or atmospheric deposition, Figure (8.1).

Sulphate (SO_4^{-2}) is found in almost all natural waters. Most sulphate compounds originate from the oxidation of sulphate ores, the presence of shales, and the existence of industrial wastes. Sulphate is one of the main dissolved constituents of rain. A high concentration of sulphate in drinking water has a laxative effect when combined with calcium and magnesium, the two most common components of water hardness.

Sulphates are compounds that are naturally present in water, due to the washing and partial dissolution of materials in the ground through which it flows (rock formations composed mainly of gypsum and sulphate soils). High concentrations have been found in both groundwater and surface water from natural sources, i.e. water that has not been subjected to anthropogenic contamination. These compounds can also appear in water through industrial wastes and discharges.

Sulphates, as they appear in drinking water, are not toxic, however in very high concentrations, a laxative effect has been observed accompanied by dehydration and gastrointestinal irritation. These waters have a bitter taste which is immediately rejected by consumers. Thus, the presence of sulphates in drinking water may cause a taste perceptible to the consumer, producing an unpleasant bitter or medicinal taste. The taste thresholds range from 250 mg/l to 1000 mg/l depending on the type of sulphate associated with sodium and calcium, respectively. Taste alteration is considered to be minimal for concentrations below 250 mg/l.

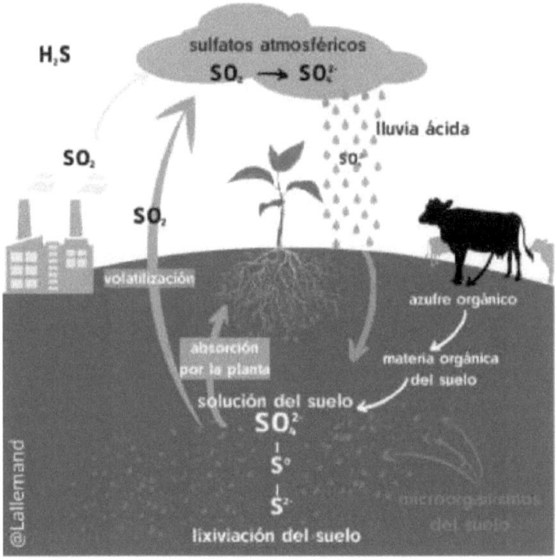

Fig. 8.1 Sulphur cycle in nature and the formation of the sulphate ion (so4).

The sulphate ion also comes from the washing of marine sedimentary materials, from the oxidation of sulphides, from the decomposition of organic substances. The dissolution of gypsum (anhydrite) provides the largest amount of sulphate to the groundwater. The sulphate ion is subject to reduction processes, especially in the presence of bacteria and organic matter.

Theoretical principle for the determination of sulphates.

The method for determining sulphates in water will definitely always be precipitation as barium sulphate. There are several volumetric proposals for its determination, however, the traditional precipitation. It will remain the most comfortable method. The sulphate ion in solution is determined by precipitation with barium chloride, in a solution weakly acidified with chlortidric acid:

$$SO_4^{-2} + BaCl_2 \rightarrow BaSO_4 + 2\,Cl^{-1}$$

The white barium sulphate precipitate exhibits physical properties such that it can be analysed by gravimetric or, where appropriate, nephelometric or turbidimetric methods.

The Kps of $BaSO_4$ at 25°C is 1.3×10^{-10} . Then the solubility of this compound, expressed in g/100 mL, can be obtained by the following calculation:

$$BaSO_4 \rightarrow Ba^{+2} + SO_4^{-2}$$

$$Kps = [Ba^{+2}][SO_4^{-2}] = s \times s = s^2$$

$$s = \sqrt{K} \quad \sqrt{1} \qquad \text{mol/L}$$

$$(1.14 \times 10^{-5}\ \text{mol/L})(233.3\ \text{g/mol})(0.1\ \text{L}/100\ \text{mL}) = 2.7 \times 10^{-4}$$

The Kps of $BaCl_2$ at 25°C is 2.9×10^{-6} . Then the solubility of this compound expressed in g/100 mL, can be calculated by:

$$BaCl_2 \rightarrow \underset{s}{Ba^{+2}} + \underset{2s}{2\,Cl^{-1}}$$

$$Kps = [Ba^{+2}][Cl^{-1}]^2 = s \times (2s)^2 = 4s^3$$

$$\sqrt{\ /\ } \quad \sqrt{\ /\ } \qquad \text{mol/L}$$

$$(8.98 \times 10^{-3}\ \text{mol/L})(208.2\ \text{g/mol})(0.1\ \text{L}/100\ \text{ml}) = 0.186\ \text{g}$$

The higher insolubility of $BaSO_4$ is observed.

Let us assume that a water sample contains 100 ppm of sulphate (SO_4^{-2}), i.e. 100 mg/L. This would represent approximately a 0.0014 M solution of SO4-2. This would represent approximately a 0.0014 M solution of SO_4^{-2}. Then, since the Kps of BaSO4 is 1.3×10^{-10} , one can calculate the grams of BaC12 that can be added to 50 mL of a 0.0014 M solution of SO_4^{-2} to initiate precipitation. BaC12 is a soluble salt that leaves Ba ions in solution^{+2} . Precipitation would occur as follows:

$$SO_4^{-2} + Ba^{+2} \rightarrow BaSO_4\ \text{(blanco)}$$

$$Pc = [SO_4^{-2}]\,[Ba^{+2}] = Kps$$

(white)

The dissociation of barium sulphate:

$$BaSO_4 \leftrightarrow Ba^{+2} + SO_4^{-2}$$

Barium sulphate will start to precipitate when:

$$[Ba^{+2}]\,[SO_4^{-2}] = 1.3 \times 10^{-10}$$

$$[Ba^{+2}] = 1.3 \times 10^{-10} / 0.0014 = 9.28 \times 10^{-8}$$

$$(9.28 \times 10^{-8})\,(208.2) = 1.93 \times 10^{-5}\,(0.05) = 9.65 \times 10^{-7}\text{g}$$

The method for the determination of sulphates consists of precipitating sulphate ions in an acid medium in the form of barium sulphate:

$$SO_4^{-2} + BaCl_2 \rightarrow BaSO_4\,(blanco) + 2\,Cl^{-1}$$

This milky-white precipitate can be analysed turbidimetrically, with scales previously established by some commercial brands, using glass tubes and nephelometric reference marks. This method would be much quicker than the gravimetric method, by weighing the barium sulphate obtained. It may be less accurate, but the realisation time is much shorter.

Material:

1 250 mL beaker

1 glass tube with scale for sulphate measurement (it is proposed to use the HI 38000 sulphate test kit. This tube can measure concentrations from 10 to 100 ppm).

Reagents:

Conditioning solution: 50 mL of glycerine is mixed with a solution containing 30 mL of concentrated hydrochloric acid, 300 mL of distilled water, 100 mL of 95% ethyl alcohol and 75 g of NaCl.

Barium chloride (hydrated salt can be used).

Procedure:

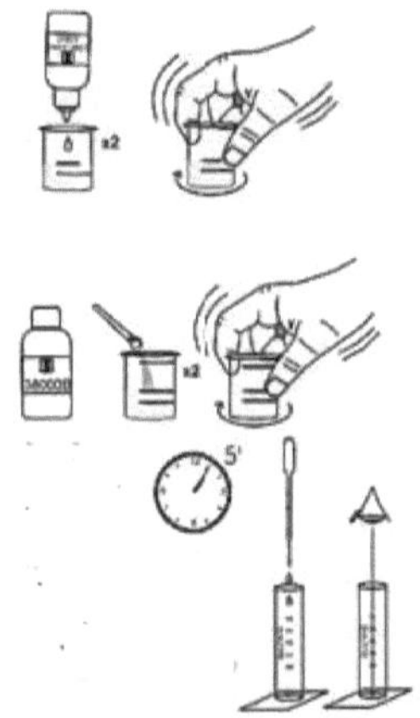

In a 250 mL beaker, place 50 mL of water sample and add about 3 drops of the conditioning solution. Shake and add a small "pinch" of barium chloride in crystals. Stir the mixture until all the added barium chloride is dissolved. The solution will turn milky-white due to the formation of the barium sulphate precipitate. Allow to stand for about five minutes. Place the measuring tube in a vertical position and add the sample solution drop by drop until the mark on the bottom of the tube is no longer visible. The direct reading in the tube is then taken, which gives an approximate indication of the amount of sulphate in the sample.

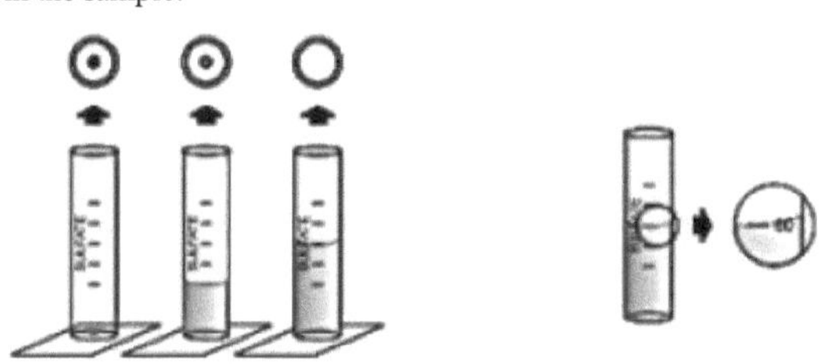

The reading shall be read directly into the tube and the sulphate concentration in mg/L is given.

Sulphates = Direct reading on the tube

CHAPTER 9

Determination of sodium and potassium

Sodium and potassium in natural waters.

Sodium is an abundant and widely disseminated element in many soils and rocks, and is the main cation of many natural waters associated with clay sediments, marine shales, evaporites and in seawater. All sodium salts are very soluble in water, so sodium-containing waters are commonly found. Sodium concentrations in natural waters vary from less than 1 mg/L in flowing waters, from areas of high rainfall to more than 100 000 mg/L in surface and groundwater associated with halite deposits in arid areas. In addition to natural sources of sodium, wastewater, industrial effluents, oilfield brines, and de-icing salts can contribute sodium to surface and groundwater. Its concentration in freshwater can vary between 1 and 150 ppm. NOM-127-SSA-1994 establishes a limit of up to 200 ppm. Sodium in drinking water can be harmful for people suffering from heart, kidney and circulatory diseases and for women with toxemias of pregnancy. High concentrations of sodium are toxic to most plants.

Potassium is found in nature in ionic or molecular form. It is a very reactive element that reacts vigorously with oxygen and water. Potassium in water is closely related to sodium and is sometimes analysed together. It comes from the weathering of feldspars and the dissolution of evaporite deposits. Although potassium is only slightly less common than sodium in igneous rocks and is more abundant in sedimentary rocks, the concentration of potassium in most natural waters is much lower than the concentration of sodium (90/10 ratio is very common). Potassium is released from silicate minerals with more difficulty than sodium and is more easily absorbed by clay minerals and reincorporated into solid weathering products. Potassium concentrations higher than 20 mg/L are unusual in natural waters. Its concentration can vary between 0.1 and 10 ppm. In brackish or warm spring waters it can be higher than 100 ppm. Large concentrations of potassium in drinking water can act as a purgative, but the range of potassium concentrations in most domestic supplies rarely causes such problems.

Determination of sodium + potassium (Na^+ + K).$^+$

The recommended methods for the determination of sodium and potassium involve the use of atomic absorption spectrophotometers or flame emission photometers. This makes these determinations very expensive. However, for the normal practice of natural water analysis and, above all, when field determinations are carried out, the procedure can be used as calculated sodium + potassium. The above expression refers to calculated values and not actually determined.

Alkaline elements are usually reported as Na + K, by the difference of the concentration of anions and cations. The calculated value, despite its limitations, is useful for many purposes in the determination of water quality. The calculation is made by determining the other most representative cations and anions present in the water in milliequivalents per litre (meq/L). The value of the difference between the sum of the anions in meq/L and the sum of the cations in meq/L is assumed to be equal to the amount of sodium plus potassium present in the water and is obtained as follows:

$$\mathbf{Na^+ + K^+ = \sum aniones - \sum cationes}$$

Most representative anions and cations in natural waters:

Anions	Cations
Bicarbonate $HCO3^{-1}$	Calcium Ca+2
Sulphates $SO4^{-2}$	Magnesium Mg+2
Chlorides Cl^{-1}	

These determinations are normally given in ppm or mg/L. For direct conversion from mg/L to meq/L the following factors can be used:

The main lions	Factor
$Ca+^2$	0.04990
Mg+2	0.0822

HCO_3^{-1}	0.0164
SO_4^{-2}	0.02082
Cl^{-1}	0.0282

Example 1: A sample of spring water has the following chemical composition:

Chemical species	Concentration in mg/L
Ca^{+2}	92.18
Mg+2	14.58
HCO_3 '1	233.
SO_4^{-2}	82
Cl^{-1}	15.75

With these data, calculate the sodium and potassium concentration by the calculated method (Na + K). First the unit conversion from mg/L to meq/L is done. Then the meq/L of cations and the meq/L of anions are added together. The difference between them corresponds to the (Na + K). To calculate the concentration of sodium and potassium, a ratio of 90/10 is established.

Cations (meq/L)	Anions (meq/L)
Ca^{+2} = 92.18 x 0.04990 = 4.59	HCO_3^{-1} = 233x 0.0164 = 3.82
Mg^{+2} = 14.58 x 0.0822 = 1.19	SO_4^{-2} = 82 x 0.02082 = 1.71
	Cl^{-1} = 15.75 x 0.0282 = 0.44
Σ = 5.78	Σ = 5.97

$[Na^+ + K^+]$ = 5.97 - 5.78 = 0.19 meq/L

To calculate the ratio of sodium and potassium considering the ratio 90/10:

Na^{+1} = (0.19)(0.9) = 0.171/0.0435 = 3.93 mg/L

K^{+1} = (0.19)(0.10) = 0.019 /0.025 = 0.76 mg/L

Example 2: A water sample has the following composition:

$[Ca^{+2}]$ = 80.16 mg/L

$[Mg^{+2}]$ = 4.03 mg/L

$[HCO_3^{-1}]$ = 233.0 mg/L

$[SO_4^{-2}]$ = 25.0 mg/L

$[Cl^{-1}]$ = 2.0 mg/L

Determine the concentration of sodium and potassium in the water sample:

Cations (meq/L)	Anions (meq/L)
Ca^{+2} = 80.16 x 0.04990 = 3.99	HCO_3^{-1} = 233x 0.0164 = 3.82
Mg^{+2} = 4.03 x 0.0822 = 0.33	SO_4^{-2} = 25 x 0.02082 = 0.52
	Cl^{-1} = 2 x 0.0282 = 0.0.056
Σ = 4.32	Σ = 4.39

$[Na^+ + K^+]$ = 4.39 - 4.32 = 0.07 meq/L

The ratio of sodium to potassium is calculated considering the ratio 90/10:

Na^{+1} = (0.07)(0.9) = 0.7733/0.0435 = 3.38 mg/L

K^{+1} = (0.07)(0.1) = 0.007/0.025 = 0.28 mg/L

The most commonly measured under field conditions are: temperature, pH, electrical conductivity; as well as micro-scale chemical analyses for HCO_3^{-1}, Cl^{-1}, SO_4^{-2}, Ca^{+2} and Mg^{+2}. The alkaline elements are usually reported as Na^+ + K^+, by means of the electrical difference of anions and cations. In addition, a replicate of the same sample can be preserved and taken for photometry.

CHAPTER 10

Anion and cation balance in a chemical analysis.

Chemical units for reporting an analysis.

Substances dissolved in water can be in molecular or ionic form, but in groundwater the ionic form is the most important. Usually these are simple ions such as the cations Na^{+}, Ca^{+2}, Fe^{+2} or simple anions such as Cl^{-1}, SO_4^{-2}. Less frequently, there are complex ions, e.g. Fe or other heavy metals, formed at the expense of organic substances or even carbonaceous acid; these complexes can be either anionic or cationic. Some of the dissolved substances are in non-ionic molecular form, often in equilibrium with ionic species. Thus, the carbonaceous acid maintains the equilibrium:

$$H_2CO_3 \leftrightarrow HCO_3^{-1} + H^{+}$$

Other substances are partially dissociated, giving rise to ions such as HCO_3^{-1}, SH^{-1}, $H_3SiO_4^{-1}$, etc., in equilibrium, with the molecular form and the fully dissociated form, or with a higher degree of dissociation. In natural groundwaters classified as fresh, most of the dissolved substances are fully ionised (the main exception is silica in the form of (H4SiO4), and are therefore considered to be in the ionic state.

Expression of concentrations

The concentration of different ions and dissolved substances can be expressed in different ways:

Part per million (ppm). It is the gram in a million grams or mg/kg. It is the most common way of expressing the concentration of a very dilute solution.

Milligram per 1 litre (mg/L). Means per litre of solution. If the total salt concentration does not exceed 5000 ppm, the density of the water is very approximately 1.0 and therefore 1 ppm = 1 mg/I.

For higher total concentrations the density is somewhat higher than 1.0 and the measurement in ppm is somewhat lower than in mg/1. However, the errors are small, even for seawater.

In chemical expressions the weights of dissolved substances are replaced by the number of moles or the number of equivalents.

$$\textit{Number of moles} = \frac{\textit{substance weight}}{\textit{molecular weight}}$$

$$\textit{Number of equivalents} = \frac{\textit{substance weight}}{\textit{equivalent weight}}$$

$$\textit{Equivalent weight} = \frac{\textit{molecular weight}}{\textit{number of electrons o Valencia}}$$

Milliequivalents per litre of solution (meq/L). It is widely used in chemical analysis because it allows direct comparison of ions. The number of equivalents per litre is called normality.

For example, to determine the meq/L of a solution containing 150 ppm (mg/L) of SO_4^{-2} the following is done:

Molecular weight of SO_4 ion^{-2} : 32 + (16 x 4) = 96

Equivalent weight of the ion (it has valence two): 96/2 = 48

150 ppm = 150 mg/L, 150/48 = 3.12 meq/L

Calculate the meq/L of a water sample containing 280 ppm Ca :^{*2}

Molecular weight of Ca ion^{+2} : 40

Equivalent weight: 40/2 = 20

280/20 = 14 meq/L

Below is a table with direct factors to convert a concentration in ppm or mg/L to meq/L of the major ions in natural waters:

Substance	Molecular weight	Factor

HCO_3^{-1}	61	0.0164
SO_4^{-2}	96	0.0208
Cl^{-1}	35.345	0.028
Ca^{+2}	40	0.05
Mg^{+2}	24.3	0.0823
Na^{+1}	23	0.0434
K^{+1}	39.1	0.0255

Table 10.1 Conversion factors for chemical units. Conversion from ppm to meq/L.

Chemical analyses have very different forms and contents depending on their intended use. Often they consist of only one or a few determinations. If the most important chemical properties of a water are to be determined, the basic ions and a few other characteristics are usually determined. Thus, the so-called complete analysis of a water -1-1-2
contains the determination of alkalinity such as HCO_3^- , Cl^- , SO_4^- , Na^{+1} and K^{+1} ,Ca^{+2} and Mg^{+2} or the sum of both as hardness. In addition, conductivity and/or dry residue, pH, organic matter, NO_3^{-1} , SiO2 and if necessary CO_3^{-2} and Fe are usually determined, although the latter are generally not important in terms of quantity. If necessary, further determinations of other ions or characteristics can be added.
In a complete chemical analysis it must be verified that: the sum of milliequivalents of anions = the sum of milliequivalents of cations:

$$r(HCO_3^{-2} + SO_4^{-2} + Cl^{-1} + NO_3^{-2}) = r(Na^{+1} + K^{+1} + Ca^{+2} + Mg^{+2})$$

where r = meq/L. Although the K^{+1} and NO_3^{-1} can in general be neglected if the water is not polluted. In practice there is a difference between the two figures, which is due to the accumulated errors of the individual determinations and not taking into account minor ionic contributions. If there is a very large difference, it can only be due to the presence of abnormal amounts of minor ions or a serious analytical error. The precision of the chemical analysis is controlled by various methods, including the difference in milliequivalents per litre between anions and cations, which must not exceed a certain value:

$$error\ (\%) = \frac{\sum aniones - \sum cationes}{\sum aniones + \sum cationes} x\ 200$$

When a calculated [Na + K] is determined in an analysis. These quantities are not used in the equation. Since it would give zero. The permissible error depends somewhat on the concentration and type of water, but as an indication the electrical conductivity in (pS/cm) can be considered:

pS/cm	50	200	500	2000	>2000
% error	30	10	8	4	4

In routine analyses, it is sometimes possible to admit slightly higher errors. If an analysis has a practically zero error, it is suspected that it has been fixed or that Na^{+1} and K^{+1} have been determined by difference.
Example: A water analysis shows the following results (Na + K. were determined by difference).

[HCO_3^{-1}] = 183.3 ppm [Ca^+] = 62.79 ppm
[SO_4^{-2}] = 50.0 ppm [Mg^{+2}] = 3.24 ppm
[Cl^{-1}] = 2.0 ppm

6Is the analysis correct? Without considering [Na + K], the meq/L of each chemical species analysed is calculated:

$[HCO_3^{-1}] = (183.3)(0.0164) = 3.0$ meq/L
$[SO_4^{-2}] = (50.0)(0.0208) = 1.04$ meq/L
$[Cl^{-1}] = (2.0)(0.028) = 0.056$ meq/L
$\sum$ aniones = 4.096 meq/L

$[Ca^{+2}] = (62.79)(0.05) = 3.14$ meq/L
$[Mg^{+2}] = (3.24)(0.0823) = 0.267$ meq/L
$\sum$ cationes = 3.407 meq/L

$$\%\,error = \frac{4.096 - 3.407}{4.096 + 3.407}(200) = 18.36$$

As the error rate exceeds 10%, it is likely that there is an error in some determination. However, it is not very far from the 10% allowed. And it can be assumed that the analysis is within an acceptable range.

At present, there are computer programmes for the treatment of hydrochemical data; among them we can mention: EASYQUIM, DIAGRAMS, SAPHIQ ,
ACUACHEM, PHREEQC and AqQA which will be discussed in the next chapter. These programmes, in their algorithm, also establish a criterion to determine whether the chemical analysis is correct or not.

For the previous example, once [Na + K] has been calculated by anion and cation difference and the data requested by the program have been entered, the following result is obtained:

▼ **Internal Consistency**

Prim a IT Tests		
Anion-Cation Balance		
Anions	4.01	
Cations	4.11	
% Difference	1.204	OK

When you see OK in the programme, it means that the analysis is correct.

When a chemical analysis shows a lot of error, a NOT will appear in the programme, as shown below:

▼ **Internal Consistency**

Primary Tests	
Anion-Cation Balance	4.88
Anions	10.1
Cations	34.890 Not within = 2% Not within = 2% Not within = 2% Not
% Difference	within = 2% Not within = 2% Not within = 2%

Example: A sample of water has the following chemical analysis. Check the accuracy of the results.

$[HCO_3^{-1}] = 220$ ppm	$[Ca^{+}] = 82.2$ ppm
$[SO_4^{-2}] = 98.3$ ppm	$[Mg^{+2}] = 17.9$ ppm
$[Cl^{-1}] = 78.0$ ppm	$[Na^{+1}] = 46.4$ ppm
$[NO_3^{-1}] = 25.6$ ppm	$[K^{+1}] = 15.5$ ppm

$[HCO_3^{-1}] = (220)(0.0164) = 3.608$
$[SO_4^{-2}] = (98.3)(0.0208) = 2.044$
$[Cl^{-1}] = (78.0)(0.028) = 2.184$
$[NO_3^{-1}] = (25.6)(0.016) = 0.409$
$\sum$ aniones = 8.245 meq/L

$[Ca^{+2}] = (82.2)(0.05) = 4.101$
$[Mg^{+2}] = (17.9)(0.0823) = 1.473$
$[Na^{+1}] = (46.4)(0.0434) = 2.013$
$[K^{+1}] = (15.5)(0.0255) = 0.395$
$\sum$ cationes = 7.982

$$\%\,error = \frac{8.245 - 7.982}{8.245 + 7.892}(200) = 3.25$$

With the AqQA programme:

▾ Internal Consistency

Primary Tests

Anion-Cation Balance

Anions	7.84	
Cations	7.98	
% Difference	0.923	OK

Measured TDS = Calculated TDS

Example: A water sample shows the following results, the [Na+K] were calculated by difference.

[HCO_3^{-1}] = 53 ppm, [Cl^{-1}] = 8 ppm, [SO_4^{-2}] = 39 ppm,
[Ca^{+2}] = 57 ppm, [Mg^{+2}] = 22 ppm [Na+K] = 26 ppm

Conversión a meq/L:

[HCO_3^{-1}] = 0.87	[Ca^{+2}] = 2.6
[SO_4^{-2}] = 1.09	[Mg^{+2}] = 1.81
[Cl^{-1}] = 0.22	

$$\% \, error = \frac{2.18 - 4.41}{2.18 + 4.41}(200) = -67.68$$

Using the AqQA programme:

▾ Internal Consistency

Primary Tests

Anion-Cation Balance

Anions	1.9	
Cations	5.47	
% Difference	48.451	Not within ± 0.2meq/L

When there is suspicion about a test result and especially when the programme gives an unsatisfactory result. The best thing to do is to repeat the analysis. If the anion and cation balance is not sufficiently accurate, the problem may also be analytical or a constituent is missing.

In most of the analyses carried out in surface waters such as streams, springs, stalactite drips, it has always been found that once the results are in ppm, when converted to meq/L, the sum of anions is always greater than the sum of cations, which means that the lack of positive charges for charge neutrality is mainly due to the sum of [Na + K].

CHAPTER 11

Graphical representation of natural water analysis.

Water analysis data can be displayed in graphical form to facilitate visual comparison. The graphical representation of hydrochemical data constitutes a very efficient working tool in the interpretation of water properties, as well as for making comparisons. It also makes it easy to see the behaviour and evolution of a water in a given territory and over time. Among the most commonly used graphical methods are the following: bar diagram, circle diagram, Stiff diagram, triangular Piper diagram, and Schoeller diagram.

Bar diagram

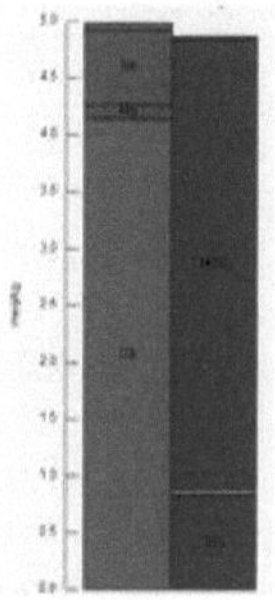

Bar or column diagrams have been widely used because of their simplicity. They were probably the first to be used. The chemical composition can be expressed in mg/L, meq/L or % meq/L. The most common form is to present in the right column the percentages of the milliequivalents of anions: $rHCO_3^{-1}$, rCl^{-1} , rSO_4^{-2} from top to bottom and in the left column: rK^{+} , rNa^{+} , rMg^{+2} , rCa^{+2} .

Circular diagram

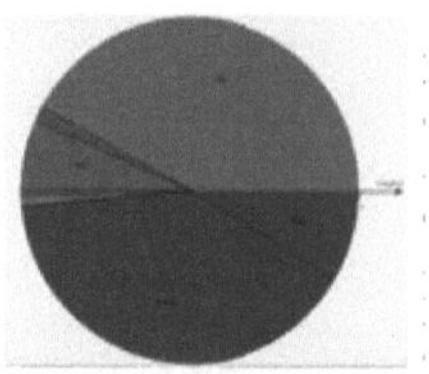

The circle diagram expresses the composition by means of a circle, whose angles are proportional to the concentrations and its radii or diameters to the total dissolved solids, the sum of the cations and anions is equal to 180°. It is a useful method for expressing the composition on the map of an area.

Stiff diagram

The Stiff diagram (1951) uses a system of parallel horizontal axes and a vertical axis. On each of these a particular ion is placed. A suitable way is to place the meq/L concentrations of the ions rMg^{+2} , rCa^{+2} , and $rNa+K^{+1}$ on the left axes from top to bottom and, in the same order on the right axes, the ions rSO_4^{-2} , $rHCO_3^{-1}$, and rCl^{-1} from top to bottom. This method makes it possible to quickly appreciate and compare the different types of water when they are found in limited quantities. It is especially useful when one wants to appreciate changes in the behaviour of a water over time due to climatic, hydrogeological or anthropogenic effects (Fig. 11.1).

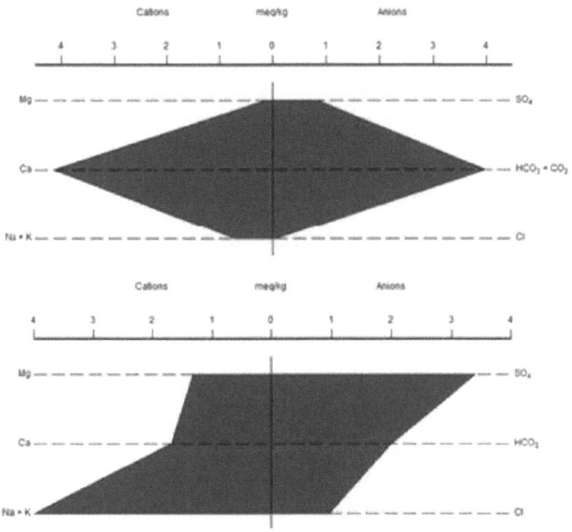

Fig. 11.1 Stiff diagrams for groundwater samples from the Sierra Gorda de Queretaro (top), and the southern part of the State (bottom).

Piper diagram

This diagram is also a graphical representation of the chemical analysis, grouping the water into families that are named after the major ions that are in higher proportion in the sample. It consists of two triangles, one for cations and one for anions, and finally a diamond-shaped figure located in the centre (Fig. 11.2).

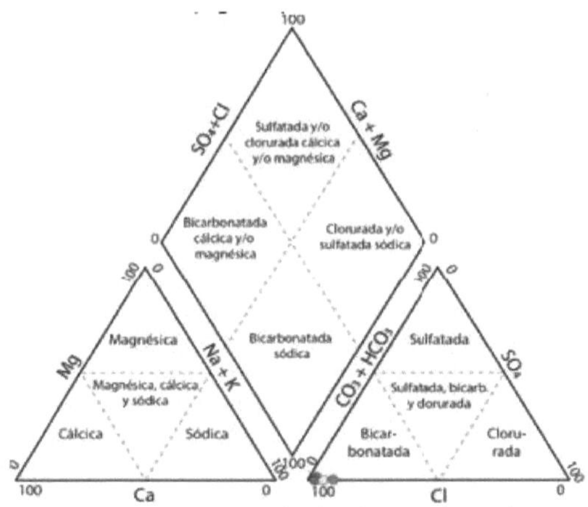

Fig. 11.2 Piper diagram, for the classification of groundwater.

The sides of the triangles are used to give the proportions of the ions. The values are shown as percentages of the total anions and cations or in meq/L. Thus, each water analysis results in one point in the cation triangle on the left and one point in the anion triangle on the right. These points are projected diagonally upwards, parallel to the diagonals of the triangles and a new point is formed on the diamond at the intersection of the two. These diagrams are used to classify waters according to their position on the diamond. When samples are taken from the same aquifer for a certain period of time, it is possible to establish the evolution that the water is following.

Schoeller diagrams.

These are so-called vertical diagrams for the graphical representation of the chemical composition of a group of samples. In these diagrams the concentration (in meq/L, or % meq/L) is placed on the ordinate axis and on the abscissa axis the different ions present. The most commonly used is the variant proposed by Schoeller (Fig. 11.3) in which the chemical composition is expressed in logarithmic units.

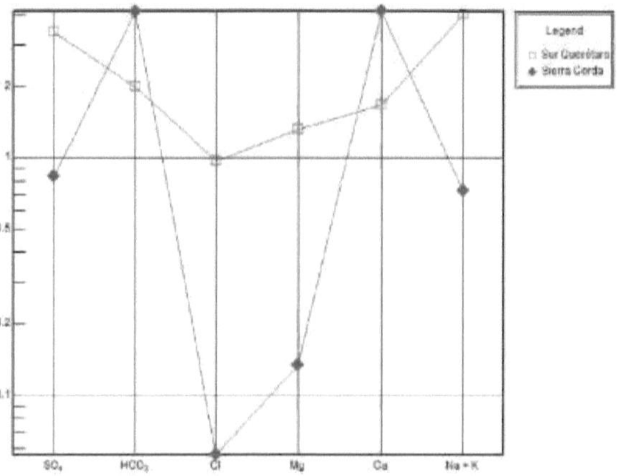

Fig. 11.3 Schoeller diagram for water samples from the southern part of Queretaro and the Sierra Gorda State.

This diagram uses an arrangement of the ions in equally spaced parallel vertical lines on logarithmic scales (in ppm or meq/L). The points are joined by a broken line. It is an appropriate diagram to show chemical similarities of groundwater. It is also suitable for studying the temporal evolution of water at the same point.

By using a semi-logarithmic scale to represent ion concentrations, with the advantage that several samples can be displayed. It is suitable for studying the temporal evolution of waters at the same site and their compositional variation at different locations.

Several parallel vertical rays or columns are arranged equally spaced and divided on a logarithmic scale and with the same modulus. An anion or a cation is associated with each ray, except the first column which has no ion associated with it and its unit of measurement is in meq/L.

Rock Were AqQA Programme

It is a very user friendly and easy to use software for hydrochemical data processing. I suggest this programme, as I have been using it for many years. To download the programme you have to buy a reserved use licence. With this programme it is possible to produce Piper and Stiff diagrams and 8 other types of hydrochemical diagrams. It performs automatic conversion of units. Checks the internal consistency of water analyses under certain criteria. Manage water data in a spreadsheet. It is a program created for anyone who uses water chemistry data. It calculates fluid properties such as water type, TDS, hardness, conductivity, saturation indices. The analysis data is entered into a spreadsheet and with a single click on the data analysis tab we check the consistency of our analysis and many water properties are automatically calculated. With another click on the new graphs tab 11 different types of hydrographic diagrams can be created automatically.

In the following I present the simple procedure of this programme for the evaluation of two water samples: one corresponds to groundwater from the northern part of the state of Queretaro and the other to the southern part or valley of the state. The results of the analysis of these two waters are presented below:

Sample A-1 (Water from the San Joaquin area, Qro.)

Sample A-2 (Groundwater from the southern part of the Queretaro Valley).

Sample A-1		Sample A-2	
Temperature	20.9	Temperature	24.7
pH	7.9	pH	8.7
ORP	27 mV	ORP	-5.20 mV
EC	500 pS/cm	EC	1217 pS/cm
SDT	249 ppm	SDT	769 ppm
Hardness	264.3 ppm	Hardness	
$HCO3^{-1}$	233.9 ppm	HCO_3^{-1}	396.5 ppm
$SO4^{-2}$	80 ppm	SO4-2	50 ppm
Cl^{-1}	20.3 ppm	Cl^{-1}	72.5 ppm
Ca+2	79.94 ppm	Ca+2	2.40 ppm
Mg+2	18.03 ppm	Mg+2	1.80 ppm
Na+	13.54 ppm	Na+	232.7ppm
K+	2.63 ppm	K+	3.80 ppm

We are going to use the AqQA programme step by step, commenting on some observations that may be useful. This programme is very easy to use and the diagrams presented, as well as the calculated information it offers, are of very good quality.

1 Once the licence for the software is installed, you will see the permission to use the software and the type of licence.

2 The program sends us to an Excel sheet to include the ppm, mg/L or mg/kg analysis.

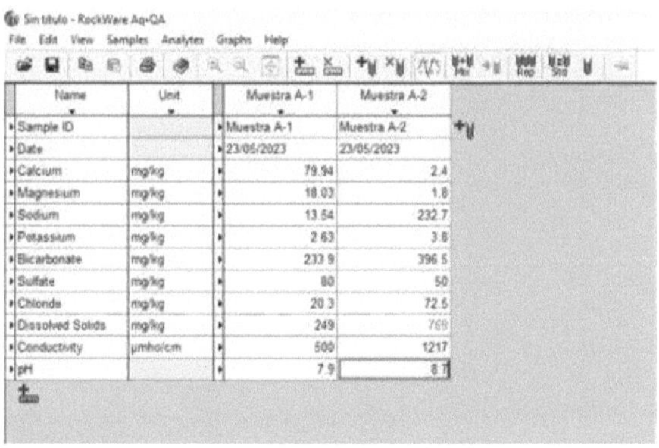

3.- Click on the Data Analisys tab and you will see: properties of the water analysed, internal consistency that evaluates the analyses carried out, carbonate balance, type of water for irrigation, and geothermometers.

Sample A-1:

Sin título - RockWare Aq-QA

Sample Muestra A-1

▾ Fluid Properties

Water Type	Ca-HCO_3		
Dissolved Solids	249 mg/kg	248.31 mg/L	Measured
Density	0.99722 g/cm^3		Calculated
Conductivity	500 µmho/cm		Measured
Hardness (as $CaCO_3$)			
Total	273.86 mg/kg	273.1 mg/L	Calculated
Carbonate	273.86	273.1	
Non-Carbonate	0.0	0.0	

▾ Internal Consistency

Primary Tests		
Anion-Cation Balance		
Anions	5.97	
Cations	6.11	
% Difference	1.169	OK
Measured TDS = Calculated TDS		
Measured	249 000	
Calculated	448 340	
Ratio	0.555	Not within range 1.0 to 1.2
Measured EC = Calculated EC		
Measured	500 000	
Calculated	565 958	
Ratio	0.883	Not within range 0.9 to 1.1
Secondary Tests		
Measured EC and Ion Sums:		
Anions	1.194157	Not within preferred range (0.9-1.1)
Cations	1.222411	Not within preferred range (0.9-1.1)
Calculated TDS to EC ratio	0.897	Not within preferred range (0.55-0.7)
Measured TDS to EC ratio	0.498	Not within preferred range (0.55-0.7)
Organic Mass Balance		
DOC ≥ Sum of Organics		
DOC unavailable		

▾ Carbonate Equilibria

Speciation at pH 7.900		
CO_3	0.01735 mmolal	
HCO_3	3.715	
CO_2	0.1023	
Total	3.834 from measured HCO_3	
Total Carbonate from Titration Alkalinity		
No Measurement for Alkalinity		
Titration Alkalinity from Total Carbonate		
	187.6 mg/kg $CaCO_3$	
Mineral Saturation		
Calcite	0.8466	Supersaturated
Aragonite	0.6823	Supersaturated
Partial Pressure of CO_2		
	0.002788 atm	

▾ Irrigation Waters

Salinity Hazard	Medium
Sodium Adsorption Ratio	356×10^{-3}
Exchangeable Sodium Ratio	0.108
Magnesium Hazard	27.1
Residual Sodium Carbonate	0.0 meq/L

Muestra A-2:

Sin título - RockWare Aq-QA

Sample Muestra A-2

▾ Fluid Properties

Water Type	Na-HCO_3		
Dissolved Solids	769 mg/kg	767.16 mg/L	Measured
Density	0.99761 g/cm^3		Calculated
Conductivity	1217 µmho/cm		Measured
Hardness (as $CaCO_3$)			
Total	13.405 mg/kg	13.373 mg/L	Calculated
Carbonate	13.405	13.373	
Non-Carbonate	0.0	0.0	

▾ Internal Consistency

Primary Tests		
Anion-Cation Balance		
Anions	9.73	
Cations	10.5	
% Difference	3.637	Not within ± 2%
Measured TDS = Calculated TDS		
Measured	769.000	
Calculated	759.700	
Ratio	1.012	OK
Measured EC = Calculated EC		
Measured	1217.000	
Calculated	866.032	
Ratio	1.405	Not within range 0.9 to 1.1
Secondary Tests		
Measured EC and Ion Sums:		
Anions	0.799303	Not within preferred range (0.9-1.1)
Cations	0.859647	Not within preferred range (0.9-1.1)
Calculated TDS to EC ratio	0.624	OK
Measured TDS to EC ratio	0.632	OK
Organic Mass Balance		
DOC ≥ Sum of Organics		
DOC unavailable		

▾ Carbonate Equilibria

Speciation at pH 8.700		
CO_3	0.1885 mmolal	
HCO_3	6.287	
CO_2	0.02729	
Total	6.503 from measured HCO_3	
Total Carbonate from Titration Alkalinity		
No Measurement for Alkalinity		
Titration Alkalinity from Total Carbonate		
	333.3 mg/kg $CaCO_3$	
Mineral Saturation		
Calcite	0.3412	Supersaturated
Aragonite	0.1768	Supersaturated
Partial Pressure of CO_2		
	743.7×10^{-6} atm	

▾ Irrigation Waters

Salinity Hazard	High
Sodium Adsorption Ratio	27.6
Exchangeable Sodium Ratio	37.785
Magnesium Hazard	55.3
Residual Sodium Carbonate	6.38 meq/L

3.- Click on the New Graph tab and a menu of 11 different hydrographic diagrams appears in a small window:

a) Plot Series
b) Time series Plot
c)Cross Plot
d) Ternary Diagram
e)Durov Diagram
f)Schoeller Diagram
g) Stiff Diagram
h) Ion Balance Diagram
i) Pie Chart and Radial Plot

Sieries Plot and Time Series Plot:

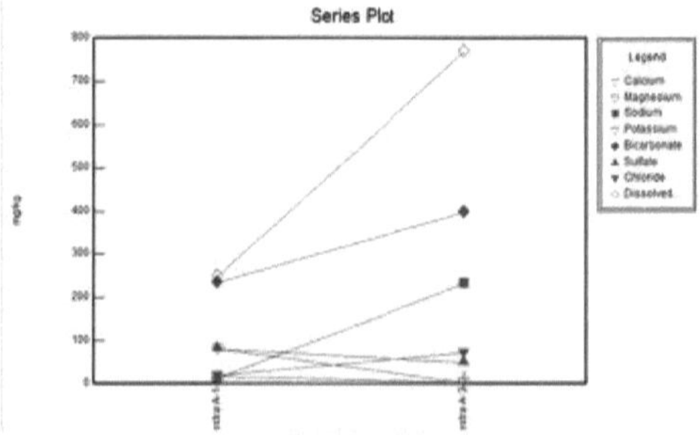

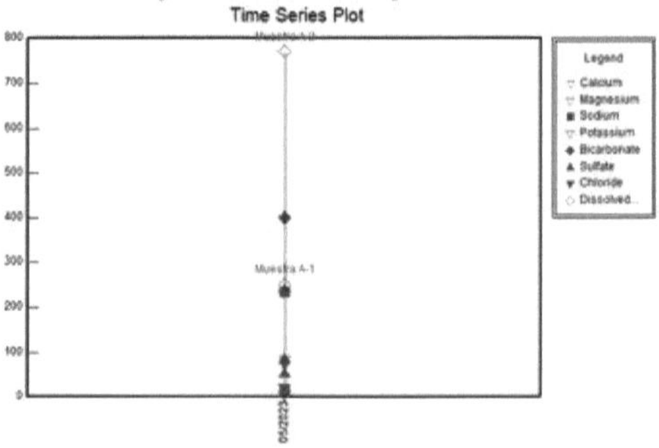

Ternary Diagram:

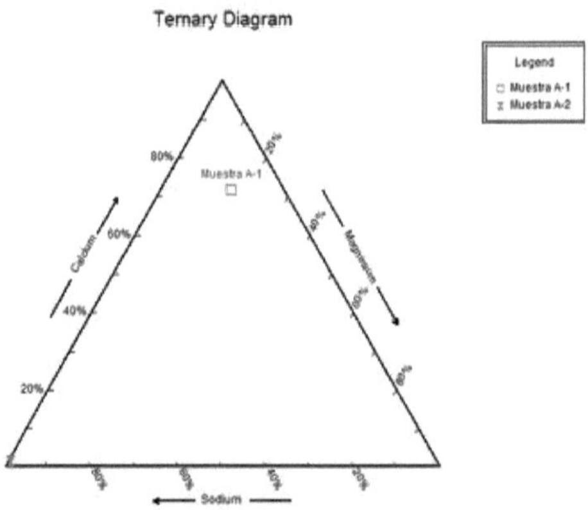

Piper Diagram:

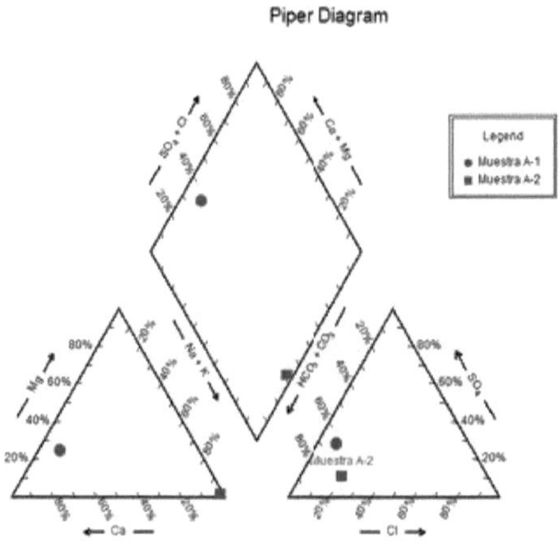

Durov Diagram:

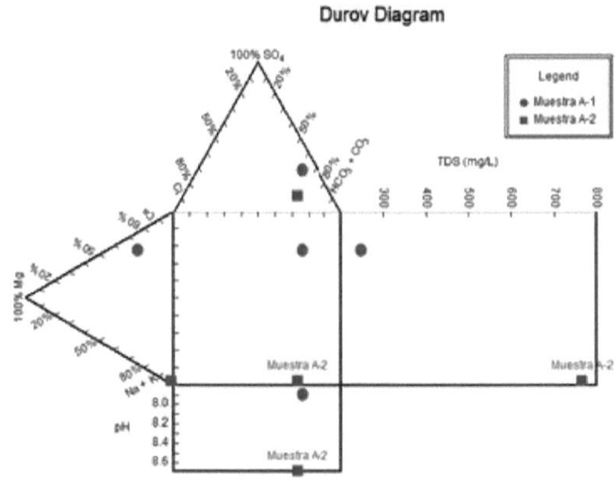

Schoeller Diagram:

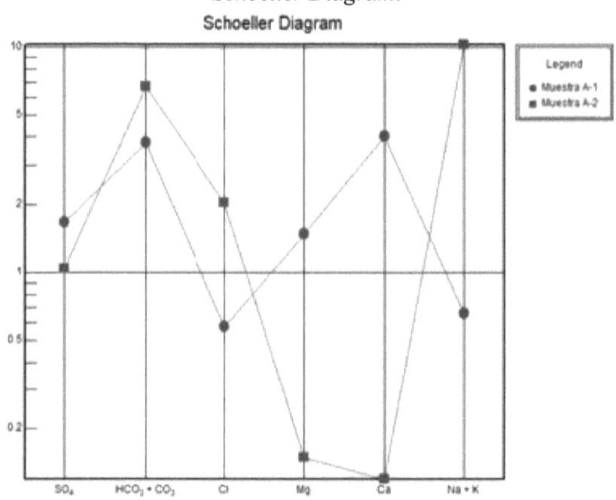

Stiff Diagram:

Stiff Diagram: Muestra A-1

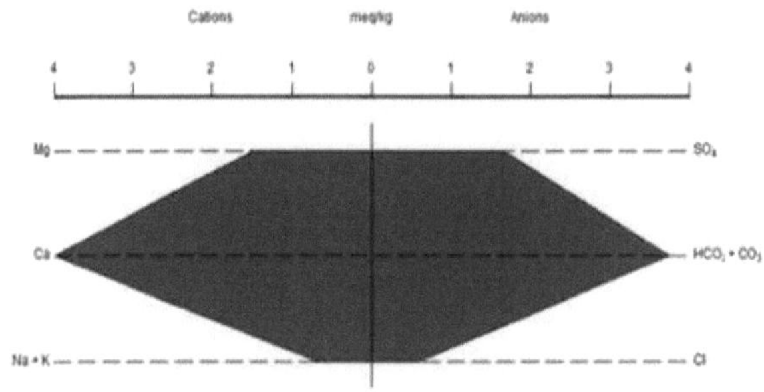

Stiff Diagram: Muestra A-2

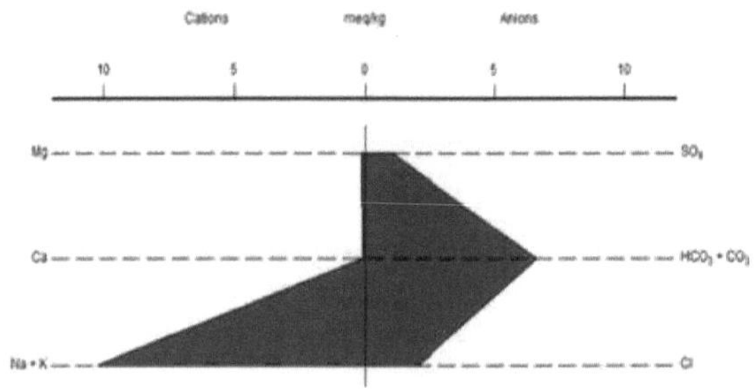

Ion Balance Diagram:

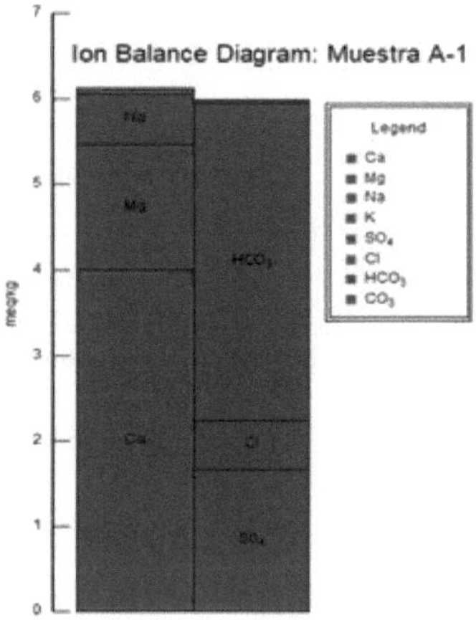

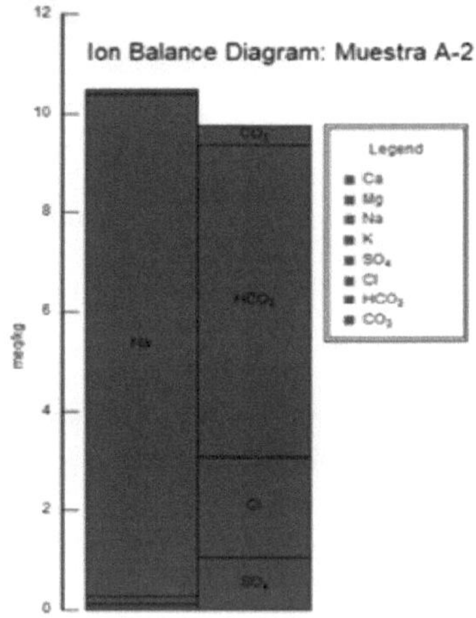

Pie Chart:

:

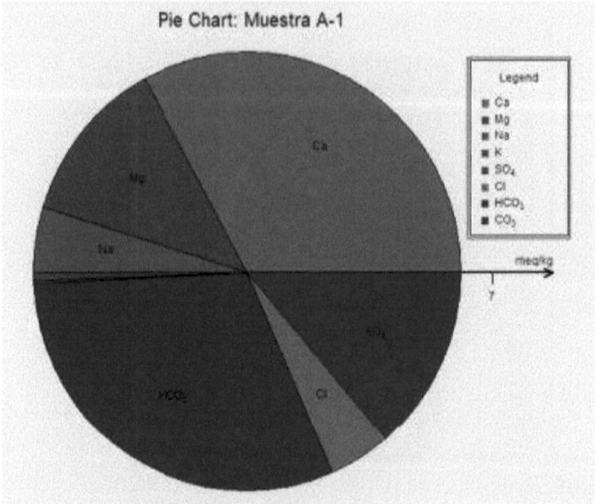

Pie Chart: Muestra A-1
Legend
Ca
Mg
Na
K
SO_4
Cl
HCO_3
CO_3
meq/kg
7

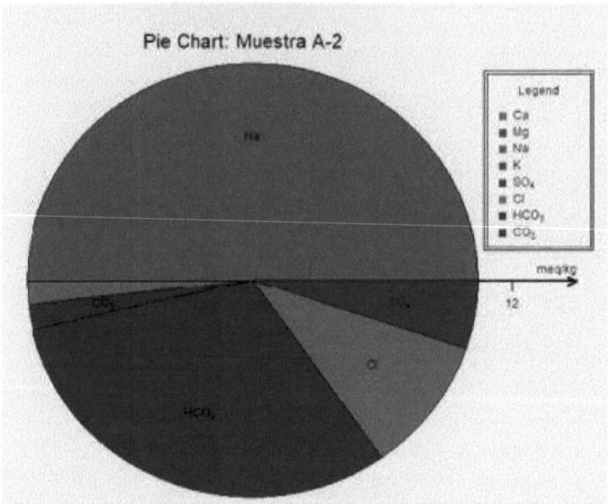

Pie Chart: Muestra A-2
Legend
Ca
Mg
Na
K
SO_4
Cl
HCO_3
CO_3
meq/kg
12

Radial Plot:

Radial Plot: Muestra A-1

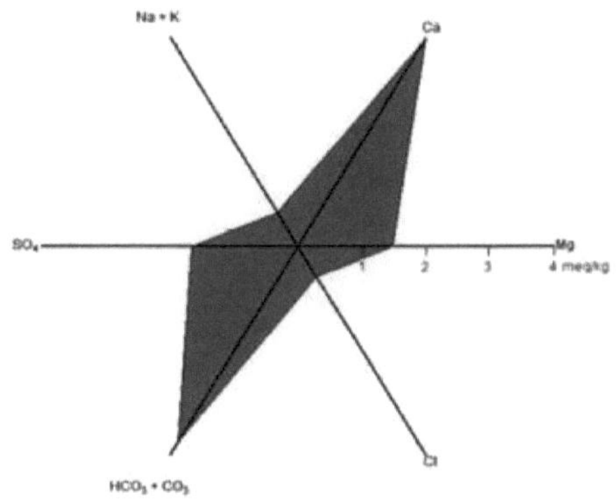

Radial Plot: Muestra A-2

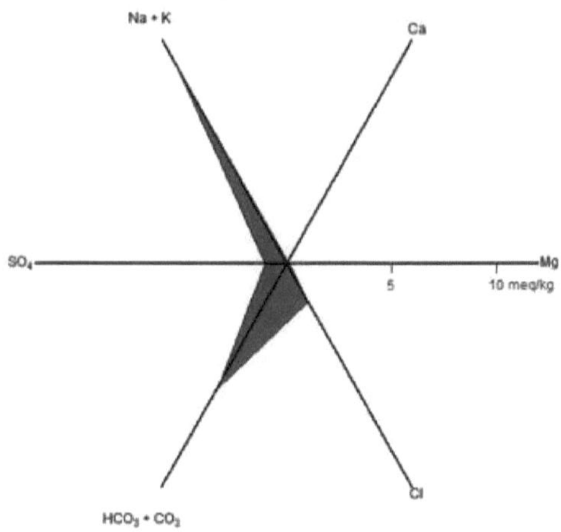

CHAPTER 12

Microscale and basic volumetric equipment.

Working with microscale.

As there is increasing concern for the environment, as well as for safety and the high costs of running a laboratory, there is a need to reduce the scale of experimental procedures. For a long time, laboratory procedures have been carried out using quantities of reagents in the order of 5 to 50 g and solvent volumes between 25 and 500 ml, known as macroscale techniques. Nowadays, there is a tendency to drastically reduce the quantities of substances to be handled, so that this methodology is being progressively displaced towards a series of less consuming methodologies and chemical steps. However, in research laboratories, considerable quantities of products are still prepared when the need arises for starting material, either for the development of a specific project or for scaling up to obtain a substance.

Most of the procedures for chemical analysis are nowadays carried out at the macroscale. This methodology allows the isolation and characterisation of the products using conventional laboratory equipment and allows very manageable quantities of products to be obtained. The main characteristics of these techniques are as follows:

- **Reagents**: Reagent quantities between 1 and 5 g are generally used. At this scale, balances, for example, with an accuracy of one tenth of a gram can be used and the errors made will not be significant.
- **Solvents**: The volumes of solvents used, both for reactions and extraction processes, are usually around 25 ml. The amount of solvent is usually not critical to the performance of the experiment. For example, if 20 ml of diethyl tier is needed for a reaction, the reaction will probably be carried out with the same yield if 18 or 25 ml are used. For the measurement of volumes, a 25 ml graduated cylinder or a 10 ml pipette will be sufficient, but this does not mean that it is not necessary to be accurate in the measurements.
- **Material**: The material used will be conventional. It will only have to be adapted to the volume of solvent or quantity of reagents used in each specific experiment.

The main objective of microscale work is to reduce the conventional laboratory scale by far:

- **Reagents**: The procedures that are carried out to

micro-weighing is carried out with amounts of the main reagent in the range 0.005 to 0.5 g. Weighing should be carried out on balances with at least two decimal places, preferably with three decimal places. Note that with these quantities, a deviation of 0.1 g in a reagent is a very significant percentage error in the appropriate proportions of the reagents used.

- **Solvents**: Solvent quantities are usually below 100 microlitres and 5 millilitres. Pipettes, micropipettes, dosing devices or syringes with the appropriate graduation and precision for each experiment should be used.
- **Material**: The material used in microscale requires adaptation to the quantities used, especially when these are less than 100 mg. This material can be of various configurations, ranging from material similar to conventional material, but with a size adapted to the needs of the quantities and volumes used in this technique, or material of a specific design (microscale *kits*).

The advantages of the use of microscale analysis techniques are very evident, among the most relevant we can cite the following:

- **It reduces costs** for each procedure.
- **It makes it possible to** increase the number and repertoire of experimental marches with the same budget.
- **It allows experiments** involving the use of more **expensive reagents**.
- **Improves safety** in the laboratory by reducing exposure to potentially toxic substances and the risk of explosion or fire.
- Significantly **reduces**

the **amount** of **reagents** used **and** consequently the **waste** generated.

- It usually involves **less** reaction and experimentation **time**, so that more time can be spent on analysing the results.
- **Improving** the **use** of **laboratories**.
- It allows the **development** of **new techniques** for the use of laboratory equipment.
- Requires **less storage space** for reagents and materials.
- **Promotes** the principle of the three

R: Reduce, Recycle, Recover.

- **It improves** the analyst's **formation**, as it forces them to be more careful at all stages of their internship work.

Nowadays, with the development of new and modern laboratory equipment, the cost of a water analysis is also increasing. Today a complete water analysis is between $1500.00 and $9800.00. Similarly, the price of reagents has risen so much (1 g of murexide is about $800.00) that doing a water analysis today requires a considerable investment. For the hydrologist, chemist, explorer, speleologist, environmentalist, analyst and anyone interested in water and environmental issues, this situation means that in very few cases are groundwater studies, tracings, drip studies, upwellings and springs carried out. Unless these studies are financed by groups and institutions that have sufficient infrastructure to finance this type of analysis.

The aim of this manual is to provide the analyst with a series of techniques, called microscale techniques, for the chemical analysis of macrocomponents of natural waters. It implements very simple techniques, notably reducing the use of reagents and, with it, the generation of waste, reducing exposure to harmful substances and increasing safety, reducing the space needed for the tests and they can also be carried out "in situ", they are techniques that reduce the cost of materials and reagents (for example one litre of EDTA 0.01 M) can be used to carry out more than 1000 (one thousand) hardness tests), they shorten the time of delivery of results.

The methodology presented here is implemented for the purpose of sampling and analysis in lakes, rivers, streams, springs, drips, caves. The use of microscale techniques allowed me to take the equipment to the bottom of some caves to perform the analysis inside them, as well as to achieve quick results of water analysis up to the area where we located the most remote springs. In short, microscale water analysis techniques save time, money and effort.

The basic volumetric equipment.

The **volumetric flask** is a flat-bottomed, pear-shaped vessel with a long thin neck (Fig. 12.1). A thin line around the neck indicates the volume contained in the flask at a normal temperature of 20 °C.

When filling measuring vessels with fluid, the following precautions should be taken:

1 Take the flask only by the neck and only at the top of the flask at the flush line.

2 Fill the flask to the mark (stripe) in such a way that the lower edge of the concave meniscus meets the stripe engraved on the neck of the flask.

3 To check that the flask is correctly filled, the observer's eyes should be at the level of the mark.

Volumetric flasks are basically used for the preparation of solutions of exact concentration. Never keep prepared solutions in these flasks for a long time, heat them or carry out any reactions in them. Flasks are available in 1000, 500, 250 and 100 mL capacity.

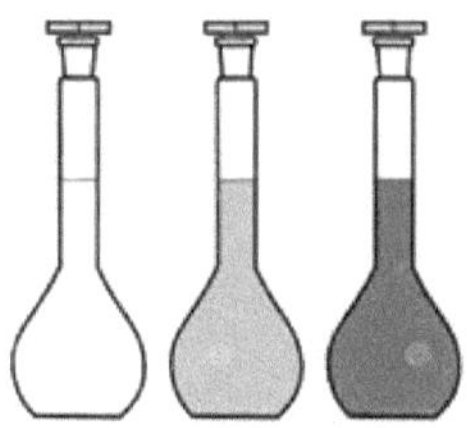

Fig. 12.1 Volumetric flasks, used for the preparation of titrated solutions.

Pipettes

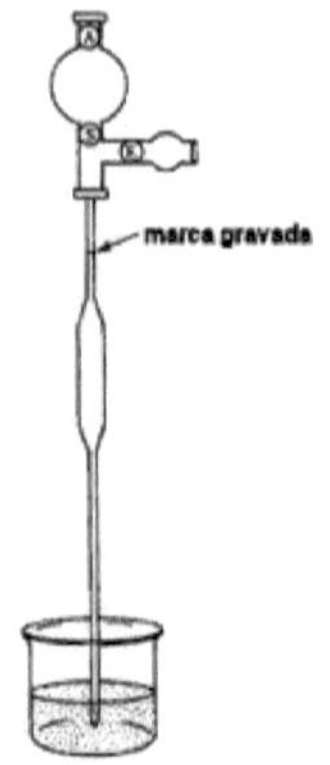

To measure smaller volumes, **pipettes** are used, which are narrow glass tubes, stretched at one end and with a widening in the middle (Fig. 12.2). Pipettes with a single mark (flush signal) are used for the exact measurement of the volume of the draining fluid. If such a pipette is filled precisely to the mark and then the liquid is poured out, the volume of liquid poured out will be equal to the volume indicated on the pipette. The most commonly used pipettes are 10 mL, and for the microscale more than 1 and 2 mL are used.

Fig. 12.2 3 mL volumetric pipette, widely used in microscale water sampling.

The volumes taken with pipettes are relatively small, therefore, errors in volume measurement with the help of pipettes can reach very large values and lead to false results. When performing volume measurement by means of a pipette, it is necessary to consider the following:

1 Work only with an absolutely clean pipette.

2 When aspirating air from the pipette with the mouth, the lower end of the pipette must always be immersed in the fluid.

3 When the pipette is filled to above the fill level, quickly remove the pipette tube from the mouth and close the opening with your index finger.

4 Holding the pipette covered with the finger in such a way that the mark is at eye level, slightly reduce the pressure of the finger on the orifice of the pipette and let the fluid pour slowly, drop by drop, until its level drops to the stop signal. Then, by squeezing the finger, stop the fluid from flowing out.

5 When all the liquid has run out, remove your finger and touch the tip of the pipette to the inner surface of the wall of the container. The pipettes are calibrated for run-off, and therefore the remaining drop of liquid retained by capillary forces cannot be blown or squeezed out of the pipette tip.

6 Observe the whole pipette very carefully and if even a single drop of undrained solution is found on the inside of the pipette, repeat the whole procedure again. Before storing a pipette, it should be washed with water and its upper orifice plugged with a paper sleeve.

7 It is advisable to use a suction knob as shown in Fig. 12.3 to avoid possible accidents when dealing with hazardous reagents.

Supplementary material

The practice of microscale water analysis also requires certain equipment for the collection and transfer of reagents for different purposes. Such as small spatulas, syringes, funnels, Pasteur pipettes, barium pipettes and Erlenmeyer flasks of 10 and 25 mL capacity (Fig. 12.3).

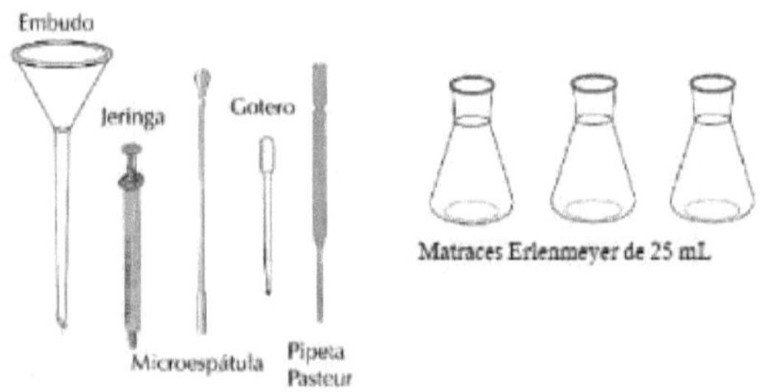

Fig. 12.3 Supplementary material for laboratory work with microscales.

Microburettes.

Of course, the most important laboratory equipment for volumetric analysis is the burette for the final titrations. In the case of microscale, the microburette is the essential part of the analytical work. This microburette allows titrations to be carried out using up to 3 mL of sample and with a consumption of up to 1 mL of standard titrant solutions. It is a small device that replaces the traditional 25 and 50 mL burettes. To manufacture a microburette, the following steps are taken:

- Pipettes graduated in hundredths of a mL (1/100) are normally used, the most usual is the 2 mL pipette, but 2, 5 and 10 mL pipettes can also be used. In this manual we recommend the 1 mL pipette.

- Lubricate a 2 cm piece of rubber tubing with a little water and unwrap it a little more than halfway up the top of the pipette.
- Place a disposable syringe without needle on the free end of the rubber hose, touching the pipette (3 to 5 mL syringe is recommended, according to the volume of the pipette. For a 1 mL pipette, the 3 mL syringe is recommended). With this syringe, by pulling the plunger, the liquid rises up the pipette to the volume mark or to the required volume and by carefully squeezing, a well controlled drip is achieved.
- Attach a plastic dispensing tip to the lower end of the pipette, so that the drops added are as small as possible, and the volume can be recorded in hundredths of a mL (e.g. a 1 mL pipette without tip produces about 15 to 20 drops and with the tip, it produces about 55 to 60 drops).

The basic microburette is shown in Fig 12.4.

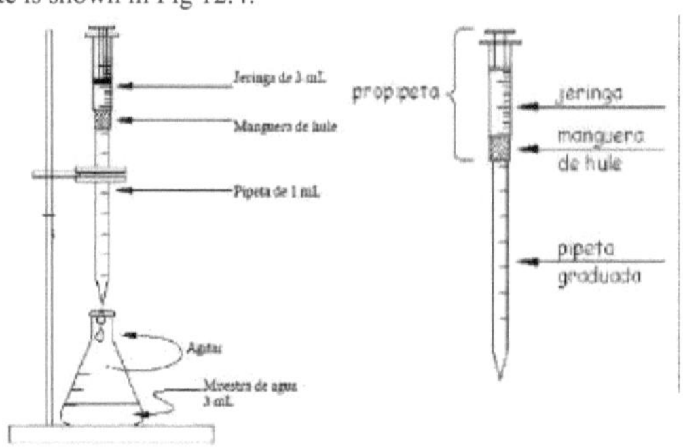

Fig. 12.4 1 mL micro burette for the microscale and assembly of the titration apparatus for 3 mL natural water samples.

Annexes

Table 1. Chemical properties of hard and soft river water.

Components	Calcareous waters	Igneous waters
SDT	Many	Few
pH	7-9	6-8
Cations	Ca^{+2} , Mg^{+2}	Na+,K+, Mg^{+2}
Anions	HCO3' $HCO3^{1}$	$HCO3'^{1}$, $HSIO3'^{1}$
Suspended solids	none	clays

Table 2. Typical groundwater analysis.

Components	Concentration (mg/L)
pH	7.2
Free CO_2	38
Total hardness (CaCO $)_3$	303
Magnesium	1.2
Calcium	119
Sodium	6.2
Potassium	1.2
Sulphate	2.5
Chloride	6.2

Extract from NOM-127-SSA1-2017

JOSE ALONSO NOVELO BAEZA, Comisionado Federal para la Protección contra Riesgos Sanitarios y Presidente del Comite Consultivo Nacional de Normalization de Regulation y Fomento Sanitario, con fundamento en lo dispuesto por los articulos 39 de la Ley Organica de la Administración Publica Federal; 4 de la Ley Federal de Procedimiento Administrativo; 3o, fraction XIII, 13, paragraph A, fraction I, 17 bis, fractions II and III, 116, 118, fraction II and 119, fraction II of the General Health Law; 38, fraction II, 40, fraction I, 43 and 47, fraction I of the Federal Law on Metrology and Standardization; 28 and 33 of the Regulations of the Federal Law on Metrology and Standardization; 209 to 213, 214, sections I, II, III and V, 215 to 225 and 227 of the Regulations of the General Law of Health on the Sanitary Control of Activities, Establishments, Products and Services; as well as 3, sections I, paragraphs n), o) and s), and II, 10, section IV of the Regulations of the Federal Commission for Protection against Sanitary Risks; I have had the honour to order the publication in the Official Gazette of the Federation on the

**DRAFT OFFICIAL MEXICAN STANDARD PRO-
NOM-127-SSA1-2017, WATER FOR DOMESTIC AND DOMESTIC USE
HUMAN CONSUMPTION. L^ PERMISSIBLE
WATER QUALITY LIMITS**

The present draft is published so that interested parties, within 60 calendar days following the date of its publication in the Diario Oficial de la Federacion, may submit their comments in writing, in Spanish and with the corresponding technical support, to the Comite Consultivo Nacional de Normalizacion de Regulacion y Fomento Sanitario, located at Oklahoma No. 14, ground floor, colonia Napoles, Alcana Benito Juarez, Codigo Postal 03810, Mexico City, telephone 50805200, extension 1333 , e-mail rfs@cofepris.gob.mx.

During the aforementioned period and in accordance with the provisions of articles 45 and 47, fraction I of the Federal Law on Metrology and Standardisation, the documents that served as the basis for the elaboration of the Project and its Regulatory Impact Analysis will be available for consultation by the general public at the domicile of the aforementioned Committee.

0. Introduction

The supply of water for human use and consumption of adequate quality is essential to prevent and avoid the transmission of water-related diseases, for which it is necessary to establish and keep up to date the permissible limits in terms of physical, chemical, microbiological and radioactive characteristics, in order to ensure and preserve the quality of water delivered to the consumer by public and private water supply systems.

For these reasons, the Ministry of Health proposes the issuance of this Mexican Official Standard, with the aim of establishing an effective sanitary control of water that undergoes potabilisation treatments in order to make it suitable for human use and consumption, in accordance with current needs.

1. Objective and field of application

1.1 This Standard establishes the permissible quality limits that water for human use and consumption must comply with.

1.2 This Standard is compulsory in the national territory for the bodies responsible for public and private water supply systems.

1.3 This standard does not apply to treated wastewater.

3. Terms and definitions

For the purposes of this Standard, the following terms and definitions apply:

3.1 Water for human use and consumption means water which does not cause adverse health effects and which does not exhibit objectionable properties or contaminants in concentrations outside the permissible limits and which does not originate from treated waste water.

3.2 Waste water means water of various compositions from discharges from public, urban, domestic, industrial, commercial, service, agricultural, livestock, treatment plants and, in general, from any use, as well as the mixture thereof.

3.3 Surface water means water that flows over the surface of the ground through streams, canals and rivers, or is stored in lakes, reservoirs, whether natural or artificial.

3.12 Responsible body, the body in charge of operating, maintaining and managing the water supply system in order to ensure and preserve the quality of the water delivered to the consumer by public and private water supply systems, establishing an effective sanitary control of the water by subjecting it to potabilisation treatments in order to make it and keep it fit for human use and consumption.

3.13 Potabilisation, the set of physical, chemical and biological operations and processes applied to water in water supply systems in order to make it fit for human use and consumption.

3.14 Water supply system, the intercommunicated or interconnected set of sources, catchment works, water treatment plants, storage and regulation tanks, conduction and distribution lines, including tanker vehicles that supply water for human use and consumption, whether publicly or privately owned.

5. Sanitary specifications

Water for human use and consumption from water supply systems must meet the following specifications:

5.1 Water in water supply systems must not be supplied from treated wastewater.

5.2 F^sicas:

Table 1 - Physical health specifications

Parameters	Permissible limit	Units
Turbidity a	4.0	UNT
pH	6.5 a 8.5	pH units
True Colour	15	UC

a The permissible limit for Turbidity will be 3.0 UNT from 2019.

5.3 Chemicals:

Table 2 - Chemical health specifications

Parameters	Permissible limit	Units
Total cyanides	0.07	mg/L
Total hardness as CaCO3	500.00	mg/L
Fluorides as F^- a	1.50	mg/L
Ammonium nitrogen	0.50	mg/L
Nitrate nitrogen (N-NO3 $)^-$	11.00	mg/L
Nitrite nitrogen (N-NO2 $)^-$	0.90	mg/L
Total dissolved solids	1000.00	mg/L
Sulphates (SO4=)	400.00	mg/L
Methylene blue active substances	0.50	mg/L

a The permissible limit for fluorides shall be 1.50 mg/L for all locations and shall be adjusted in accordance with the phased-in compliance table Table 3 of

.4 Metals and metalloids:

Table 4 - Metal and metalloid health specifications

Parameters	Permissible limit	Units
Aluminium	0.20	mg/L
Arsenic a	0.025	mg/L
Barium	1.3	mg/L
Cadmium [b]	0.005	mg/L
Copper	2.00	mg/L
Total chromium	0.05	mg/L
Iron	0.30	mg/L
Manganese	0.15	mg/L
Mercury	0.006	mg/L
Nickel	0.07	mg/L
Lead	0.01	mg/L
Selenium	0.04	mg/L

NOTE 1 Permissible limits for metals and metalloids refer to their total concentration in water, which includes suspended and dissolved metals and metalloids.

[a] The allowable limit for arsenic shall be 0.025 mg/L for all locations and shall be adjusted in accordance with the phase-in compliance table Table 5 of this section 5.4.

[b] The permissible limit for cadmium shall be 0.005 mg/L for all locations and shall be adjusted in accordance with the phase-in compliance table Table 5 of this section 5.4.

Preparation of standard solutions used in this manual.

0.01M AgNO3 solution. Approximately 2 g of silver nitrate are weighed and taken to the oven where it is left for one hour at 110°C. After this time it is allowed to cool in a desiccator. For the 0.01 M solution, exactly 1.699 g of AgNO3 is weighed out and dissolved with water, preferably bi-distilled water to make up to 1 L. As the action of light affects the silver nitrate, it is advisable to keep the solution in amber glass bottles, or instead in clear glass bottles lined with dark paper.

0.01 M EDTA solution. Dry a 5 g sample of the purified dihydrate (Na2H2Y.2H2O, formula weight = 372.2) at 80°C to remove surface moisture. Then allow to cool in a desiccator and weigh 3.722 g and dissolve to 1 L with distilled water until saturated.

H2SO4 solution 0.02 N. A 0.1 N solution of this acid contains 0.4906 g of H2SO4 per L. 25 mL of a 0.02 N solution are placed in a flask and:

$$(50mL)(0.1N) = (0.02W)(x)$$

125 ml , diluted to a volume of 125 mL

Bibliography consulted

Ayres H. G. *Quantitative Chemical Analysis,* HARLA, 2ª edition, 1970.
Brumblay Ray U., *Analisis Cuantitativo*, CECSA, 1969.
Crites R., Tchobanoglous G., *Wastewater Treatment in Small Populations*, McGrawHill, 2000.
Custodio E., Llamas M.R., *Hidrologia Subterranea*, Editorial Omega, 1996.
Day, JR. R.A., Underwood A.L. *Quimica Analitica Cuantitativa*, 5ª . Edicion, Prentice-Hall, Hispanoamerica, S.A. 1986.
De Heredia Alonso J.B.,Torregrossa A.J., Gonzalez M.T., Dominguez Vargas J.R., *Analisis Quimica de Aguas Residuales.* Abecedario. University of Extremadura, 2004.
Fagundo Castillo J.R., *Hidroquimica del Karst*, Ediciones Osuna- Cuba, 1996.
Gavino de la Torre G, Juarez L.C., Figueroa T. H.H., *Tecnicas Biologicas de Laboratorio y Campo*, LIMUSA, 1999.
Geissler Gunther, Arroyo Maribel, *El Agua como un recurso natural renovable*, Ed. Trillas,2011.
Helwelg Otto J. *Recursos Hidraulicos*, LIMUSA, 1992.
Ibanez Cornejo J.G., Mainero Mancera R.M., Doria Serrano M.C., *Experimentos de Quimica en microescala para nivel medio superior*, Universidad Iberoamericana, 2009.
Jander Gerhart, *Volumetric Analysis*, UTEHA, 1961.
Jenkis David, *Quimica del Agua* (Manual de laboratorio), LIMUSA, 1983.
Juarez Clemente, R. Lemus C., *Manual de Qumica Aplicada,* Ed. ROREZ, 1966.
Kreshkov A., Yaroslavtsev A., *Course of Analytical Chemistry, Quantitative Analysis,* Mir Publishing House, 1985.
Ledesma Ledesma G. *Las Cavernas del Area de San Joaqum, Queretaro*, EXCAV/COBAQ/H. Ayuntamiento de San Joaqum, 2022.
Lopez A.P. *Abastecimiento de Agua Potable*, Alfaomega, IPN, 2002.
Lopez Cancio J.A., *Problemas de Qumica*, Prentice Hall, 2006.
Maskew F.G., Charles G.J., Alexander O.D., *Ingenieria Sanitaria y de Aguas Residuales*, Ediciones Ciencia y Tecnica S.A., Volume 1 and 3, 1987.
Mere Alcocer F.J. *Tratamiento y control de la contamination de las aguas urbanas*, CPPEQ, 2003.
NOM-A-A-73-1981, for chlorides.
NMX-AA-036-SCFI-2001, for alkalinity.
NMX-AA-074-SCFI-2014 , for the determination of sulphates.
NOM-AA-72-1981, for the determination of hardness.
Romero Rojas J.A., *Calidad del Agua*, alfaomega, 2000.
Orozco D. F. *Analisis Qumico Cuantitativo*, Editorial Porrua, 1977.
Smog Douglas A, West Donald M., *Introduction a la Qumica Analitica,* editorial reverte, s.a.1986.
Symons James M. *Mas claro, ni el agua*, CEA-Queretaro, 1997.
Vega de Kuyper J.C. *Qumica del Medio Ambiente*, Alfaomega, second edition, 2007.
Vogel Arthur I. *Quantitative Anahetic Chemistry. Volume I.*
KAPELUSZ Publishing House. 1960.

MIX
Papier aus verantwortungsvollen Quellen
Paper from responsible sources
FSC® C105338

Printed by Books on Demand GmbH, Norderstedt / Germany